Chromosomal Q-heterochromatin in the Human Genome

Chromosomal Q-heterochromatin in the Human Genome

By

Abyt Ibraimov

Cambridge
Scholars
Publishing

Chromosomal Q-heterochromatin in the Human Genome

By Abyt Ibraimov

This book first published 2020

Cambridge Scholars Publishing

Lady Stephenson Library, Newcastle upon Tyne, NE6 2PA, UK

British Library Cataloguing in Publication Data
A catalogue record for this book is available from the British Library

ISBN (10): 1-5275-5975-0
ISBN (13): 978-1-5275-5975-2

To my parents, wife, and sons

CONTENTS

ABSTRACT

The results of nearly 50 years of study into human chromosomal Q-heterochromatin regions (Q-HRs) are summed up in this book. Research in this area has not been systematized since the 1980s. The results of this research show that this type of constitutive heterochromatin is found only in the genome of the three higher primates (*Homo sapiens*, *Pan troglodytes*, and *Gorilla gorilla*). It has also been found that: a) there is wide variability in the amount of chromosomal Q-HRs, which are primarily inherent in human populations; b) interpopulation differences are undoubtedly related to environmental factors, rather than to racial or ethnic features. This detailed account presents the history of the problem, the phenomenon of differential fluorescence of chromosomes, and the methodological approaches used in their study. Data are also presented on the distribution of chromosomal Q-HRs in relation to sex, age, norm, and pathology. The bulk of existing data on human chromosomal Q-HRs testify that they constitute a self-sustaining genetic system subject to natural selection and, together with C-HRs, play a part in cell thermoregulation, depending upon the amount present in the genome of a given individual, with all its ensuing consequences.

PREFACE

Supporters of Darwinism have gathered convincing evidence that primates originated about 70 million years ago when the global climate was warmer and the current continents had a different location: North America was connected to Europe and Asia, while South America and Africa were separated from North America by a wide sea. Further evolution of primates took place against the background of serious climatic and geographical changes, for example, the global climate has gradually become cooler, especially during the winter seasons. Forest areas began to decrease, giving way to open treeless savannas and areas covered with ice appeared on the surface of the planet.

Especially dramatic changes in the global climate have occurred in the last 2-2.5 million years and these changes coincided with the appearance of the first hominids. Science does not yet know whether the emergence of the first *Homo species* was the result of profound climate change on the African continent, or whether cold weather and other physical environmental factors contributed to the emergence of some species and the extinction of others, thereby accelerating the evolution of all mammals. Since very few species of *Homo* have survived up to the present day, it is difficult to develop a full picture predicting events in the near future, including at the genome level.

It is only known that fluctuations in climate in the last 2.5 million years have had a strong impact on all organisms and ecosystems and hominids are no exception. Be that as it may, the evolutionary success of our ancestors during this period is evidence of their high plasticity, first in Africa and then in the face of the new, even more severe, natural conditions of Eurasia. However, this test was passed by only one species (*Homo sapiens*). The remaining surviving primates (about 230 species) remained in the tropical and subtropical provinces of Africa, South America, India, and Southeast Asia.

Most people believe that man was always set to occupy the dominant position on Earth—the place that he now occupies. However, evolutionary biologists believe that even the very origin of man was not an inevitable event and he has repeatedly been on the verge of extinction. Our knowledge of human biology and paleoanthropology shows that 20,000-30,000 years ago we were not even outwardly as different as we are now. Most of these external anthropological (racial) differences are considered to be the result of adaptive human evolution to different climatogeographic conditions and some of them appeared after some of our ancestors left Africa and settled Eurasia and Oceania. We are now more than seven billion in number and we have inhabited almost all the lands of the Earth, including the Far North, high-altitude areas, and hot deserts. Some 12,000 years ago, about ten million people lived on the Earth.

About a century and a half ago, Thomas Huxley, a follower and contemporary of Charles Darwin, compared the anatomical features of man with other primates and showed that there are no fundamental differences between them. His task was to establish unity between man and the animal world. The results of comparative studies conducted in recent decades at the molecular and chromosomal level have fully confirmed the correctness of Darwin and Huxley. Moreover, these data suggest that the differences between humans and chimpanzees are even slightly smaller than those between chimpanzees and gorillas. The results of studies on the hybridization of DNA between humans and chimpanzees show a commonality of almost 98.7 %—a value that allows the possibility of a viable hybrid.

The totality of the existing data shows that all modern humans are descended from one ancestral group of hominids who lived about 100,000-130,000 years ago in East Africa, and not millions of years ago as once assumed. 30,000-50,000 years ago, some of them managed to leave Africa and expanded to inhabit the whole of the Earth. However, how they managed to do so, while remaining a single tropical species, is a subject of long-standing and heated debate.

It is believed that the uniqueness of adaptive human evolution consisted of a combination of three components: high physiological plasticity, cultural evolution, and natural selection. If this is the case, then where, from what, and when did humans first appear. Man has high physiological plasticity and, most importantly, a high-functioning mind, without which there would be no cultural evolution. The question remains as to whether these factors are genetic in nature.

As is well known, man does not have a single protein or enzyme that has no analogue in the world of other living beings. However, in another respect the human genome is unique: among the animals and plants studied to date, only the genome of the three highest primates—human, chimpanzees, and gorillas—contain Q-heterochromatin. As it turns out, unlike chimpanzees and gorillas, only humans display a wide hereditary variability, the detailed description of which is the subject of this work.

After the collapse of the USSR, our laboratory received financial support in 1993 and 1995 from the International Science Foundation, USA and DIFCO Laboratories Ltd., Great Britain, for which we offer many thanks.

I express my sincere thanks to my colleagues who worked with me over the years in the Laboratory of Human Genetics at the National Center of Cardiology and Internal Medicine of the Kyrgyz Republic, including E.I. Aksenrod, S. Zhumaliev, G.U. Kurmanova, S.A. Nazarenko, G.O. Karagulova and S. Tabaldiev. Quantitative analysis and mathematical modeling were performed with the participation of E.Kh. Ginzburg (Institute of Cytology and Genetics, RAS, Novosibirsk) and V.V. Popov (Institute of Physics and Mathematics of the NAS of the Kyrgyz Republic). Although the title page of this book does not have their last names, they are as much authors of this work as I. I am responsible for the evaluation of data from other authors, without whom this monograph would not exist and, of course, for any errors.

Bishkek, 2020

PART I

THE PROBLEM

"Nothing in Biology Makes Sense Except in the Light of Evolution".
—Th. Dobzhansky (American Biology Teacher, 1973, 35: 125-129).

Over five million years of evolution, separating man from his closest kinsmen (the chimpanzee (*Pan troglodytes*) and the gorilla (*Gorilla gorilla*)), he has changed a lot: he has lost his fur, became completely upright, learned how to make tools, mastered speech, and created culture and science. Humans now number more than seven billion and we have settled, practically, the entire land area of the Earth, including the circumpolar belt, high altitude areas, and deserts (some 12,000 years ago, about ten million people lived on the Earth). All this happened relatively recently, as man only went beyond East Africa 30,000-50,000 years ago.

Inevitably, the question arises: why did all this happen? The answer to this question, it would seem, is that humans developed reason and turned north, while some of them reached Patagonia through Beringia. Few people doubt that man has succeeded thanks to his ability to maintain the tropical microenvironment around his body ("tropicalization") in the form of clothing and buildings. Prior to this, our distant ancestors lived where our closest evolutionary relatives (the chimpanzee and gorilla) remain today.

If this process is connected to the capacity of the human mind, then how did man become the owner of such a powerful advantage? What changes occurred in his body? Perhaps during this time, we acquired some unique genes or complexes of genes that even our closest relatives in evolutionary history, the chimpanzee, does not have? Molecular biology data suggests that the differences between humans and chimpanzees are even slightly smaller than the differences between chimpanzees and gorillas and the results of studies on the hybridization of DNA between humans and chimpanzees shows a commonality of almost 98.7%—a value that allows for a viable hybrid.

Perhaps the methods of modern biology are insufficient to understand it? Far from it. As such, why has only one species out of the 230 in the primate family had such an obvious advantage? Why did other mammals who arose long before primates and whose descendants exist in Africa or another continent not acquire such a gene(s)? Common sense and scientific experience accumulated since the time of Lamarck and Darwin suggest that the human genome should still have some features that, along with the cultural heritage accumulated by human history, provide for the dominant position of *Homo sapiens sapiens* on Earth.

However, the true nature of such features, including genetic ones, remains unknown. Since people from different parts of the Earth or various racial-ethnic groups produce quite normal and viable descendants, there are no serious reasons to believe that significant genetic changes have occurred in the human genome during their adaptation to different climatic and geographical provinces.

On the other hand, it is known that between individuals belonging to different racial/ethnic groups, there are indeed inherited external differences, as evidenced by children born from interethnic marriages. Since these inherited anthropological differences (races) have arisen in the last 20,000 years, and this period coincides with the history of modern human development of the most diverse climatic and geographical provinces of the Earth, we have to admit that these external differences seem to be related to this adaptive evolution.

It is considered that man was initially well adapted to a hot and dry climate. This consideration is based on two arguments: 1) man as a species was formed in the tropical and subtropical climates of Africa; 2) man has very effective physiological mechanisms to combat the overheating of his body, like sweating. To all this, man, depending on the specifics of his permanent place of residence and lifestyle, has developed many behavioral and cultural reactions. Since many other animals (e.g. the camel) are adapted to a hot and dry climate, and sometimes even better adapted than humans, it is not clear why this should be: were the same genes (or complexes of genes) inherent in all mammals used or did the adaptive evolution of man occur in a completely different fashion?

In the north, permanent human sites appeared long before the appearance of the first civilizations. Nevertheless, modern man does not well tolerate the effects of cold and, in this regard, is inferior to all those homoeothermic animals that live constantly at high geographic latitudes. The reasons for this are well known: his absence of thick dense wool, a thin layer of subcutaneous fat, and a high metabolic rate with relatively large body size, etc.

Most primates are also characterized by relatively low tolerance to cold stress, with the exception of a small colony of Japanese macaques (*Macaca fuscata*) that withstand the cold winters of Central Japan. However, their bodies are covered with thick long hair. In contrast, prior to the invention of primitive clothing, housing, and controlled fire, our ancestors lived in places where the ambient temperature did not fall below +10 °C.

Unfortunately, the surface of the Earth, unlike a cartographic map, is uneven. One third of terrestrial land is covered with mountains, sometimes reaching up to 8,000 m and more above sea level. Nevertheless, over the past few millennia man has managed to populate the lower levels of the world's mountains to an altitude of 4,200 m. As it turns out, of all the climatogeographic territories settled by man, the high altitude areas (more than 2,500 m above sea level) have proved the most severe. For endotherms, to which man belongs, high altitudes show: 1) low average annual atmospheric temperatures, which require, among other things, increased energy consumption; 2) low partial pressure of pO2, limiting locomotor activity, and heat production by reducing muscle contractions. Thus, at high altitudes man faces a double danger: he is forced to consume and spend more energy more intensively than in the conditions of lowlands and he has to do all this in conditions of constant oxygen deficiency. In addition, at high altitudes we find dry air, increased radiation, and frequent winds, etc.

In studying the genetic basis of the adaptive evolution of man, scientific approaches sufficiently tested by experience and time has not yet been developed, we are able to focus on the methodology of the general theory of evolutionary genetics. Lewontin [1974] suggests that: "The alternative natural populations to direct measurements of fitness or fitness components is to show at least that selection must be operating, even so it cannot be measured, by correlating the frequencies of alternative alleles with temporal or spatial differences in environment". In other words, in the first approximation, it is possible to rely on both the general similarity and differences in the frequencies of the studied genetic material between populations living in different climatic and geographical conditions. If populations differ significantly in the frequencies of a trait under study, we can postulate local adaptation, but if they are very similar, we can assume general adaptation.

Traditionally, the biology of adaptive human evolution has been studied at the molecular, chromosomal, and organismic levels using the full range of methods of modern medical and biological sciences. This book is devoted to the description of adaptive human evolution at the level of chromosomes and the whole organism.

PART II

THE FACTS

"Patterning of chromosomes is of far greater importance than hitherto realized, yet we have not yet found the tools permitting us to analyze this phenomenon"
—Ernst Mayr [2002].

2.1. Chromosomal heterochromatin regions (HRs) and their variability

Modern human cytogenetics is more than seventy years old. Its biography begins in 1956 when Tjio and Levan first established that the exact number of chromosomes in the human diploid set is 46. The early satisfaction found in establishing the exact number of the diploid set of chromosomes in the human karyotype and the discovery of chromosomal diseases, the so-called "trisomic period" [Lejeune et al. 1959; Ford et al. 1959; Jacobs et al. 1959], soon ran into difficulties once the need arose for accurate identification of each chromosome. The fact is that with the chromosome staining methods of that time, it was only possible to partially identify the human chromosome set. Exactly the same situation was observed in the cytogenetic study of other mammals. This was at a time when insect cytogenetics, using the analysis of polytene chromosomes, achieved impressive success and was already engaged in compiling a physical and genetic map of chromosomes based on the study of their fine morphology. Attempts to use the chronology of DNA synthesis, quantitative measurements of the length of chromosome arms, and other methods for identifying human chromosomes did not give the desired results [Therman 1986].

1968 marked the beginning of a new era in cytogenetics. A group of Swedish scientists led by Caspersson published the results of the first cytochemical studies, indicating that the pattern of differential fluorescence and the length of metaphase chromosomes could be used for full and accurate identification [Caspersson et al. 1968, 1970].

The history of this important methodological discovery certainly deserves a mention. Hsu [1979], the author of the first essay on the history of human and mammalian cytogenetics, and, by the way, one of the authors of the C-staining technique, which initiated the systematic study of heterochromatin regions of eukaryotic chromosomes [Arrighi and Hsu 1971], discussed the origin of the Q-staining technique of chromosomes. At the end of the 1960s, Prof. Torbjörn Caspersson was invited to attend the Children's Cancer Research Foundation in Boston as a consultant. It was unlikely that S. Farber, the director of this Foundation, who invited T. Caspersson, an outstanding cytologist and recognized authority in the field of fluorimetry and interferometry, knew that his choice would have such revolutionary consequences for the development of cytogenetics, particularly human cytogenetics.

By that time it was known that many fluorochromes can stain chromosomes and fluoresce under ultraviolet irradiation. However, all these fluorochromes showed a uniform fluorescence along the entire length of the chromosome. Caspersson, however, focused on whether it was possible to attach reagents to the fluorochrome molecule so that the first could connect to guanines in the chromosomes and the second give fluorescence. Additionally, if the distribution of base pairs along the chromosomes were not accidental, then those parts rich in GC pairs would bind more actively with molecular dyes than chromosomes rich in AT pairs, while detecting bright fluorescence (as well as the fact that knowing the ratio of GC/AT pairs was more important then than now). Therefore, with the validity of all these assumptions, one may expect that chromosomes under a fluorescent microscope will fluoresce differentially, revealing alternating bright and dark zones. As such, it would be possible to distinguish chromosomes by the pattern of their fluorescence. Caspersson persuaded Ed Modest, an organic chemist from the same institution, to synthesize a suitable compound—quinacrine mustard—which was sent for testing to the Karolinska Institute in Stockholm.

The first series of experiments was carried out on plant objects (*Scilla*, *Vicia*, etc.) and under a fluorescent microscope metaphase chromosomes were clearly differentiated by length. At the same time, each chromosome had its own individual alternating pattern of bright and dark, of weak or non-fluorescent segments, which made it possible to completely and accurately identify homologues [Caspersson et al. 1968, 1969a, b].

The first article by the Caspersson group presented pictures of the differential fluorescence of the chromosomes in *Vicia faba*, Chinese hamsters, and *Trillium erectum*, demonstrating the ability to accurately identify each homologous pair in the karyotype. Two years later, this group adapted a differential staining method to identify the metaphase chromosomes of the human karyotype—the Q-banding (quinacrine-based banding of chromosomes) approach introduced by Lore Zech [Caspersson et al. 1970]. Later, the Paris Conference on Standardization in Human Cytogenetics recommended this technique be named Q-staining (from English "quinacrine mustard") and chromosome segments giving fluorescence were called "Q-bands" [Paris conference 1971].

It is well known that in the use of routine (uniform) color, not all human chromosomes differ sufficiently in total length, nor in the ratio of the lengths of short and long arms, making it difficult to fully identify them. After staining with quinacrine mustard (Q-staining), human chromosomes take on the form of a transversely striated structure in which one can clearly distinguish the segments—from areas that almost non-fluoresce to those with the high (brilliant) intensity. The alternation of such differentially fluorescing segments for each homologous pair of chromosomes is so individual that it makes it possible to accurately identify any metaphase chromosome in the human karyotype. Thus, for the first time, it was possible to undertake an exact and complete identification of each individual chromosome—this was a turning point in the development of human cytogenetics. However, the possibilities of the Q-staining method of chromosomes had not yet been exhausted.

It soon became clear that the constancy of the pattern of differential fluorescence along the length of individual homologous pairs of chromosomes is not absolute. It turned out that, in the human karyotype, one or both homologues of some chromosomes can differ in the presence of particularly brightly fluorescent segments. The aspersson group noted that the distal part of the long arm of the Y chromosome, which is well expressed in the interphase nuclei, also fluoresces most clearly [Zech 1969; Caspersson et al. 1970]. The same authors showed, for the first time, the existence of variable fluorescent segments on the pericentromeric region of autosome 3 and the short arms of chromosomes 13-15. The variability of human metaphase chromosomes was noted by many other researchers on these and on the 4, 21, and 22 autosomal pairs [van der Hagen and Berg 1970; Schnedl 1971; Evans et al. 1971; Cervenka et al. 1971; Lin et al. 1971; Moscetti et al. 1971; Pearson 1972; et al.].

Before proceeding to an analysis of the data obtained by the differential staining of chromosomes, it is necessary to provide some brief information about the history of the problem of the hereditary variability (polymorphism) of chromosomes in human populations.

In the first studies on the morphology of human metaphase chromosomes using routine staining methods, it was noted that certain regions of some chromosomes varied considerably in their microscopic structure without appreciable influence on the phenotype of the carrier. Such variants were found on the long arms of chromosomes 1, 9, and 16; on the short arms of acrocentric chromosomes; and on the long arm of the Y chromosome. In particular, they manifested themselves in the form of stretched and weakly stained areas, secondary constrictions, in the pericentromeric regions of their long arms on chromosomes 1, 9, and 16. Acrocentric chromosomes varied in the presence of the proximal part of the short arm, in the length of the satellite filaments, and in the size of the satellites. Routine methods of coloring have shown the wide variability of the size of the long arm of the Y chromosome. A peculiar and rare morphological variant has been described for chromosome 17, in which secondary constrictions may appear on a short arm giving the impression of the presence of a satellite region there. Many other very rare variants of chromosomes in the human karyotype have also been described. At the same time, numerous researchers have shown that the chromosomal variants described above are stable in ontogenesis and are inherited. The first information about the existence of racial differences in the frequency of chromosome variants (the size of the long arm of the Y chromosome; the inversion of chromosome 9) were also described before the advent of modern methods of differential chromosome staining.

A detailed analysis of the literature on the routine variants of human chromosomes is not part of our task and they are mainly of historical interest. The fact is that studies of human metaphase chromosomes by methods of differential staining have shown that all chromosomes in the human karyotype are potentially polymorphic—this condition is not the exception, but the rule. The data obtained in recent years indicate an unexpectedly high variability (polymorphism) in human chromosomes and the uniqueness of the karyotype of each individual in human populations. Readers interested in studies on chromosomal polymorphism in humans prior to the introduction of modern methods for the differential staining of chromosomes can find comprehensive information in a number of special and review works [de la Chapelle 1961; de la Chapelle et al. 1963; Ferguson-Smith et al. 1962; Court Brown et al. 1965; Therkelsen et al. 1967; Priest et al. 1970; Lubs, Ruddle 1971; Zakharov 1977; Zakharov et al. 1982; Mikelsaar 1979; Prokofyeva-Belgovskaya 1986].

Any great leap in scientific development implies the emergence of new ideas or research methods. In human cytogenetics, new research methods have usually outpaced new ideas. This occurred with the discovery of the phenomenon of chromosomal polymorphism of heterochromatin regions (HRs) in the human karyotype. Arrighi and Hsu [1971] developed a fundamentally different technique for the differential staining of chromosomal HRs using Giemsa dye, now known as the C-staining technique. This staining selectively reveals one type of constitutive heterochromatin that exists on all human chromosomes. Therefore, chromosomes stained by the C-technique ("constitutive heterochromatin") are commonly called "C-bands" to emphasize the particular type of heterochromatin detected by this method [Paris Conference 1971]. At present, this technique, with small modifications, has been introduced into cytogenetic practice as one of the main methods for studying the variability of chromosomal HRs in eukaryotes, including humans.

Fluorescent polymorphism of chromosomes, differing from quinacrine polymorphism (Q-staining), reveals Hoechst 33 258 fluorochrome and DAPI (or DIPI) [Schnedl et al. 1977]. The benzimidazole derivative, Hoechst 33 258, which is especially actively fluorescent when interacting with A-T nucleotides, also stains human chromosomes unevenly [Raposa and Natarajan 1974], but the definition of differential fluorescence over the length of a chromosome is much worse. Unlike quinacrine, quinacrine-mustard, or other acridine derivatives, when stained using this technique, heterochromatin near the centromeric regions of chromosomes 1, 9, and 16, the short arm of chromosome 15, and the distal part of the long arm of Y chromosome fluoresce. However, this technique is not widely used in studying the variability of Q-HRs in human chromosomes.

A number of informative methods have been developed that reveal the polymorphism of the nucleolus-organizer regions (NORs) of human acrocentric chromosomes. Those NORs, localized in the satellite threads of all acrocentric chromosomes, can be specifically stained by 'silvering' (Ag-staining) [Goodpasture and Bloom 1975; Howell et al. 1975]. It has now been established that nucleolar organizers undergo Ag-staining if they were active in the previous interphase [Miller et al. 1977]. To date, some very important information has been obtained on the nature of Ag-polymorphic variants of chromosomes, on the nature of their inheritance and tissue differences, as well as on the population frequencies of Ag-polymorphic variants [Mikelsaar et al. 1977; Verma et al. 1981]. In addition to the good reproducibility of the method and its availability for conventional unspecialized cytogenetic laboratories, Ag polymorphism is attractive to researchers because the function of the NORs of acrocentric chromosomes is well known and cistrons for 18S and 28S ribosomal DNA are localized.

Thus, based on the data obtained by the above-mentioned methods of chromosomal differential staining, it turns out that there is much wider chromosomal polymorphism in human populations than was expected based on data obtained using routine staining before the 1970s. The results of recent studies show that there are many variants of chromosomes in populations that can be detected by various methods (differential staining techniques, *in situ* hybridization of nucleic acids, autoradiography, etc.) and the combination of these variants determines the unique karyotype of each human. Therefore, it is no longer possible to consider the human karyotype as the only morphological standard; it is clear that it is a combination of chromosomal variants, with one or another frequency occurring in populations of normal individuals [Prokofyeva-Belgovskaya 1986; Zakharov et al. 1982; Paris Conference 1971, Suppl. 1975; ISCN 1978; Verma 1988; Bhasin 2005]. Numerous studies have shown that the chromosome regions, identified by these methods as heterochromatic regions, actually possess the fundamental properties characteristic of constitutive heterochromatin.

More than ninety years ago, E. Heitz found that the substance making up the bulk of the nucleus—chromatin—consists of two components: 1) euchromatin, which contains genes, and 2) heterochromatin, which does not carry such genetic material [Heitz 1928]. To imagine the significance of this discovery, it is best to turn to the classic work itself. In 1935, in his review "The Structure of Chromosomes and Genes", based on the achievements of genetics and cytology of the time as well as his own research, Heitz [1935] came to the following conclusions:

> "(1) All chromosomes show a longitudinal differentiation into euchromatin and heterochromatin that relates to the genetic properties of each chromosome. (2) The differentiation is specific for each chromosome and is different in the karyogram of each animal and plant species. ... (5) Heterochromatin formation and the degree of chromosomal contraction are genetically determined, and heterochromatin is located at corresponding positions of homologous chromosomes. (6) Chromocenters of interphase nuclei result from equilocal positioning of heterochromatin of different chromosomes. (7) Species can be distinguished by their size and pattern of chromatin distribution. (8) Euchromatin is closely connected to gene activity during interphase; heterochromatin corresponds to genetically inert regions. ... (11) Sex chromosomes are frequently subject to heterochromatin formation" (cited in Passarge 1979).

Heitz was also the first to suggest a name for the union of cytology and genetics, then distant to each other as scientific fields: cytological genetics.

The very term "heterochromatin" is a "hybrid" of the terms "heteropicnosis" and "heterochromosomes" used to describe the behavior of sex chromosomes in mitosis in cytology. By proposing the term euchromatin, Heitz was referring to those parts of the chromosomes that are no longer visible under an optical microscope at the end of the telophase. It is not by chance that Heitz specifically clarified the word "heteropicnosis" as a term denoting the differential behavior of the whole chromosome or its part in the prophase and telophase at a certain stage or throughout the whole development cycle of an individual. The term "heterochromosome" refers to the whole chromosome, while "heterochromatin" refers only to the part of the chromosome that remains heteropicnotic in interphase.

Shortly after Heitz's death, Brown [1966] suggested the terms "constitutive" and "facultative" heterochromatin. Prokofyeva-Belgovskaya wrote that "recently it seems most correct to identify heterochromatin as specific chromosome regions whose DNA is formed by clusters of high and medium-repeating, often satellite, non-transcribed nucleotide sequences. The obligatory property of heterochromatin is the condensed state throughout the entire cell cycle, late replication of the DNA contained in it and positive staining by the C-method" [Prokofyeva-Belgovskaya 1986].

In the last collective monograph devoted to heterochromatin, one of the authors writes:

> "There was a time when it could be argued that heterochromatin differed from euchromatin in its behavior but not in its fundamental structure [Baker and Callan 1950; Dyer 1964; Brown 1966]. That time is now past. Constitutive heterochromatin is indeed composed of DNA sequences with distinctive characteristics. Moreover, there are not two classes of heterochromatin, constitutive and facultative, as is still commonly claimed. The use of the noun *heterochromatin* to describe euchromatin that is facultatively inactivated is both misleading and unnecessary. The facultative heterochromatinization of euchromatin, in principle, has more in common with tissue-specific condensation of euchromatin, though it obviously takes place on a more extended scale, both temporally and spatially, within the individual organism and usually to only one sex" [John 1988].

Time has confirmed the validity of almost all the conclusions of E. Heitz, although he may have been mistaken on a few small things, such as in the use of the characteristic of the formation of chromocenters as an important criterion for determining heterochromatin. We now know that chromocenters are not formed in all cell types and this is true for both constitutive heterochromatin and facultative heterochromatinized euchromatin [John 1988], which also does not form chromocenters. Analyzing the results obtained using a simple cytological method of "boiling" ("Kochmethode"), Heitz managed to achieve an understanding of not only the genetic nature of heterochromatin, but also anticipate one of the main methods of differential chromosome staining—C-staining. After all, in essence, C-staining [Arrighi and Hsu 1971] is a modification of the Heitz "boiling" technique (the stage of preliminary heat treatment of chromosomes before staining).

2.2. Genetic peculiarities of chromosomal HRs

Brown [1966] proposed distinguishing between two types of heterochromatin: constitutive and facultative. Two homologous chromosomes carrying constitutive heterochromatin—one from the father, the other from the mother—have the same behavior and the same morphological appearance during both embryogenesis and in adults. In the case of facultative heterochromatin (X chromosome), the two homologues behave differently: during development (in early embryogenesis), one of the two homologues becomes partially or completely heterochromatinized, while the second remains euchromatic. This difference persists in the adult body. In facultative heterochromatin, genes are in a state of temporary repression. This is due to the fact that euchromatin regions (ER) located near heterochromatin are not despiralized in the interphase nucleus and remain condensed (heterochromatinized) [Lyon 1961]. As for constitutive heterochromatin, it is recognized that it is usually not transcribed, since the DNA of these HRs has a peculiar molecular composition and they contain mostly moderately or highly repetitive nucleotide sequences.

A classic example of facultative heterochromatin is Barr's body, which is the result of the inactivation of one of the two X chromosomes of a female at an early stage of embryogenesis. It is believed that heterochromatinization is an effective mechanism aimed at ensuring that both sexes have only one genetically active X chromosome per cell [Ohno 1967]. Sex chromosomes are examples of facultative or constitutive heterochromatin, at least in insects and mammals. Thus, the Y chromosome, for most of its length, is formed by constitutive heterochromatin. The same applies to the *w* chromosome in birds.

In many species, the X chromosome makes up about 5 % of the genome—this is the case in humans, dogs, and mice, etc. It is believed that these species retained the original X chromosome, characteristic of a common ancestor [Ohno1967]. Some species of rodent have X chromosome of very large size. Ohno [1967] believes that, in the course of evolution, these large X chromosomes were formed as a result of duplications, triplications, or quadruplications fully reproducing X chromosomes. In this case, in addition to the complete inactivation of one of the X chromosomes in the female, heterochromatinization extends to the remaining euchromatin region of the male's X chromosome, as well as that of the female. As such, the "universal goal" of this heterochromatinization is the preservation of the genetic equivalent of the original X chromosome in each somatic cell. In other words, the dose of genes associated with the X chromosome, the mechanism of which was explained by Lyon in 1961, is compensated for in the female's cells. However, as Prokofyeva-Belgovskaya [1986] quite reasonably noted, the use of the terms "facultative" and "constitutive" heterochromatin [Brown 1966] is incorrect. She considered it appropriate to remove the term "facultative heterochromatin" and replace it with the term "inactivated euchromatin", which more accurately reflects the nature of the structure and its condition, i.e. in this case the ER undergoes "heterochromatinization". The essence of the phenomenon of heterochromatinization is that a portion of the ER close to the heterochromatin acquires the properties of the latter and goes into a condensed state, repressing the activity of the genes localized there. The process of heterochromatinization is reversible, labile, controlled by the genotype, and affected by developmental conditions; it may not appear in all cells, causing the mosaic development of this trait.

Taylor [1960], using radioautography to study the replication pattern of Chinese hamster chromosomes, described the late replication of the DNA of HRs: in the male a significant part of the long arm of the X and Y chromosomes synthesize DNA at the end of the S phase of the cell cycle. In the female, one of the X chromosomes has the same label as the male X chromosome, while the other only labels late. Taylor found in one species of mammal one of the most characteristic genetic properties of chromosomal HRs—late DNA replication—which had been discovered by Lima-de-Faria [1959] a year earlier in insect sex chromosomes. In *Microtus agrestis*, large blocks of heterochromatin are replicated during the last four hours of the S phase [Pera 1968].

Late replication of heterochromatin DNA is also characteristic of human chromosomal HRs and is a stable property of the X chromosome in women and the Y chromosome in men. In addition, each chromosome in the human karyotype has an area that replicates at the end of the S phase; therefore, these areas see a characteristic localization of the label on radioautographs, allowing us to identify many of them [Schmid 1963; German 1964; Ganner and Evans 1971; Schnedl 1972]. Breg et al. [1972] established that the Q+ and G+ bands mainly correspond to late replicating regions of chromosomes. These observations are consistent with Comings's [1972, 1973] description of the "intercalary" type of heterochromatin. According to the author, intercalary heterochromatin should be distributed along the length of the chromosome arms and correspond to the Q+ and

G+ bands of chromosomes. According to Comings, the DNA of intercalary heterochromatin contains only a small number of highly repetitive nucleotide sequences.

It is believed that there is a correlation between the physical state of chromatin (condensation) and its genetic activity. Regarding the giant chromosomes of Drosophila, it is known that heterochromatin does not take any part in the formation of the characteristic morphological formations in which structural genes are located. Segments of giant (polytene) chromosomes with their characteristic pattern (transverse banding) only form euchromatin regions. As for the HRs of the Drosophila chromosome, which is one-third composed of heterochromatin, it has been established that they do not participate in the formation of characteristic segments where structural genes are localized. It has long been known that the loss of a significant part of chromosomal HRs is usually not lethal, in contrast to the euchromatin material of chromosomes where the loss of even the smallest part is fatal. In addition, the duplication of extensive heterochromatin segments of chromosomes does not cause significant disruption of genetic homeostasis, which also contrasts with the serious consequences of the duplication of ERs in chromosomes.

Finally, crossing-over, the main genetic process of meiosis, usually affects only the euchromatin regions of the chromosomes. Chiasmas, which are the morphological manifestation of this process, are only exceptionally observed in the HRs of chromosomes. The absence of chiasm has been shown in species where chromosomes have extensive heterochromatin zones, for example, hedgehogs and field mice. This has been confirmed at the molecular level [Hotta and Stern 1978]. These authors showed that satellite DNA, which is an important component of heterochromatin, does not include a radioactive label at the pachytene stage. In view of the fact that this stage is characterized by residual DNA synthesis, necessary for recombination, it can be concluded that it does not take place in heterochromatin.

Radioautographic studies of DNA transcription have also revealed the genetic inertness of heterochromatin. Sieger et al. [1970] showed that in the constitutive heterochromatin nuclei of *Microtus agrestis* no radioactive label is detected after the introduction of tritiated uridine, as in the characteristic chromocenter of the nuclei of somatic cells of the quail [Mirre and Stahl 1978].

All these observations suggest that heterochromatin is a genetically inert material in eukaryotic chromosomes. However, the question still arises as to whether this this almost universally accepted hypothesis, in its absolute form, is too great a simplification of the problems arising in the study of heterochromatin? Indeed, it has long been known that the genetic activity of euchromatin may be altered or violated, and even completely inhibited, if, due to mutation, some part of the heterochromatin is located near euchromatin sites (the "position effect"). It has also been noted that the amount of heterochromatin contained in the nucleus affects the manifestation of some quantitative traits of the organism [Prokofyeva-Belgovskaya 1986; John 1988]. There are other data indicating that chromosomal HRs are far from inert genetic material, the description of which is the primary focus of this work.

2.3. Cytological peculiarities of chromosomal HRs

Currently, the existing methods of differential staining of chromosomes can identify two types of constitutive heterochromatin: C and Q. Several methods are used to identify chromosomal C-heterochromatin regions (C-HRs). The most commonly used method is C-staining [Arrighi and Hsu 1971; Sumner 1972], which makes it possible to stain pericentromeric C-HRs of chromosomes 1, 9, and 16, as well as HRs on the distal part of the long arm of the Y chromosome. The method, called "G-11" [Gagne and Laberge 1972; Bobrow et al. 1972], selectively stains C-heterochromatin in the area of secondary constriction of the long arm of chromosome 9.

The method of Q-staining is mainly used to identify chromosomal Q-HRs, [Caspersson et al. 1970]. DAPI fluorescence (4,6-diaminido-2-phenylindole), through the action of distamycin, reveals HRs in areas of secondary constriction on the long arms of autosomes 1, 9, and 16, the q12 segment of the Y chromosome, and the short arm of chromosome 15 [Miller et al. 1974].

The combination of these methods has allowed us to reach a new level in the study of chromosomal HRs. However, it should be noted that, as Prokofyeva-Belgovskaya [1986] rightly emphasized, "by the 50s our information about the microscopic structure and properties of heterochromatic regions of chromosomes was quite complete. It was obvious that these areas of chromosomes are specialized areas of chromosomes, firmly fixed in them in the process of evolution and clearly distinguished from euchromatic structures by a number of properties". One of the genetic peculiarities of chromosomal HRs has been considered above. Here, we summarize some other cytological features of HRs detected before the 1970s, i.e. before the advent of modern methods of differential staining:

1) All chromosomal HRs are capable of conjugating to each other, forming chromocenters and loops inside the chromosomes and terminal telomeric compounds. This suggests that HRs form if not identical, then biochemically very similar, loci and their conjugation should be considered true homologous conjugation [Prokofyeva-Belgovskaya 1986]. The aggregation of HRs was also observed by Heitz [I934]. He showed that the HRs of the giant Drosophila chromosomes merge to form a bulky chromocenter;
2) HRs under the influence of certain environmental factors (such as cold, illness, and hunger) can be partially despiralized and their DNA concentration decreases. Darlington and La Cour [1940], under cold conditions, were able to localize HRs in the chromosomes of many plants. The euchromatin sites of chromosomes are very resistant to changing environmental conditions;
3) Under the influence of mutagenic factors, as well as some viruses that sever the chromosome, chromosomal HRs are damaged [Natarajan and Ahnström I972];
4) HRs occupy vital areas of chromosomes: centromeres, telomeric ends of chromosomes, regions of nucleolar organizers, and most of the sex chromosomes. With the help of modern techniques of differential staining, it has been possible to confirm that the genetic, cytological, and molecular features of chromosomal heterochromatic material, previously established for different types of eukaryotes, are also characteristic of human chromosomal HRs. The application of these methods has led to the development of a series of new observations of interest.

The most significant of these has been the detection of the wide hereditary variability of chromosomal HRs (see Part II). Data were obtained on the "activity" of heterochromatin in interphase nuclei. In mice, for example, it forms dense chromatin masses located near the nuclear envelope and around the nucleolus [Rae and Francke 1972]. Using radioautography with the inclusion of tritium-labeled thymidine, Comings and Okada [1973] localized heterochromatin by its late label in *Microtus agrestis*, on the one hand, in the chromocenters corresponding to the sex chromosomes, and on the other hand, in peripheral chromatin, which, according to the authors, corresponds to intercalary heterochromatin. The peripheral localization of most of the heterochromatin in the nucleus was also proved for plant cells through the pattern of its late replication in radioautographies with particularly high resolution [Sparvoli et al. 1977].

2.4. Molecular peculiarities of chromosomal HRs

Even before the advent of modern methods of analyzing the longitudinal chemical differentiation of chromosomes (methods of *in situ* hybridization, radioautography, selective fluorescence, and immunochemistry) based on the genetic and cytological features of heterochromatin, the notion that chromosome HRs are formed by a series of numerous, if not identical, biochemically very similar loci with small, complementary effects on some quantitative characteristics of the organism, was developed [Prokofyev-Belgovskaya 1986; Mather I944]. It was postulated that the analogy of the biochemical composition of heterochromatin underlies the attraction, aggregation, and non-homologous pairing of chromosomal HRs.

In this regard, the conclusion of Britten and Kohne [1968] that the DNA of higher organisms contains highly repetitive nucleotide sequences was decisive. At the same time, many cytogeneticists suggested that this class of DNA would be concentrated mainly in chromosomal HRs. The first studies of the biochemical composition of the DNA contained in heterochromatin seemed to confirm the hypothesis of the similarity of HR DNA sequences. Fractionation of DNA isolated from euchromatin and heterochromatin during equilibrium centrifugation on hydroxylapatite, separately and in combination, highlighted the significant heterogeneity of human DNA. Such studies carried out on the cell nuclei of mice and other mammals showed that the DNA of these animals was significantly enriched by satellite DNA—a type of DNA with repetitive nucleotide sequences. Corneo et al. [1967, 1968] found that when DNA was centrifuged on a density gradient of neutral CsCl, two peaks were found corresponding to primary DNA, with a density of 1.700 g/cm3, and satellite I, with a density of 1.687 g/cm3. Yasmineh and Yunis [1973] showed that heterochromatin in mice consists of satellite DNA, and seems to be of the same type, which could, in fact, be the molecular basis of the cytological properties of HRs. However, subsequent studies have shown the improbability of this hypothesis, because it turns out that heterochromatin has a complex and heterogeneous DNA composition.

In 1970, Pardue and Gall, as well as Jones [1970], showed that satellite DNA was localized in centromere regions by *in situ* hybridization on mouse chromosome preparations. At least four types of satellite DNA (satellite I-IV) have been identified in human DNA. Satellite DNA I is present in high concentrations in the distal region of the long arm of the Y chromosome and weakly in the centromeric heterochromatin, less frequently in the telomeric regions of chromosomes 1, 3-5, 9, 12-17, and

19-22 [Jones et al. 1974]. The HRs of chromosomes 1, 3, 13, 14, 16, and 21 were the most constantly labeled. The authors note that this type of satellite DNA is localized in those segments of chromosomes that usually show bright fluorescence after staining with quinacrine.

Satellite DNA II, rich in A-T pairs of nitrogenous bases, is mainly localized in the area of secondary constriction of chromosomes 1 and 16, and, to a lesser extent, in a similar region of chromosome 9 [Jones and Corneo 1971]. A significantly smaller number of satellite II is located in the pericentromeric regions of the chromosomes of groups D, F, and G. The affiliation of satellite II to heterochromatin is proven by the fact that, in the interphase nuclei, RNA complementary to satellite II is localized mainly in chromocenters. Satellite IV is contained in the HRs of many autosomes, especially in the acrocentric and Y chromosomes [Gosden et al. 1975].

Evans et al. [1974] and Gosden et al. [1975] showed that different satellites have preferential localization in different chromosomes. At the same time, Y chromosomes are especially distinguished, containing satellite I and IV, chromosomes 1 and 16 with a predominant localization of satellite II, and chromosome 9 especially rich in satellite III. From the above, we can conclude that satellite DNA is found in all regions of chromosomes formed by constitutive heterochromatin. Miklos and John [1979] noted that in humans, satellite DNA is only 4 % of the genome, while the total amount of constitutive heterochromatin detected by the C-staining method is approximately 20 % of the genome. Thus, it should be recognized that heterochromatin consists not only of satellite DNA. Analysis of human DNA using restriction enzymes revealed the existence of a wide variety of highly repetitive sequences that are not satellites. Manuelidis [1978] was able to show that a DNA fragment of 340 base pairs, isolated by the restriction enzyme *Eco RI*, is concentrated in C-segments of chromosomes, and satellites I-IV (chromosomes 7, 10, and 19) were not found in the heterochromatin of chromosome 1.

New research has suggested that human heterochromatin is heterogeneous in its molecular composition and simultaneously contains satellite and other types of DNA with highly repetitive nucleotide sequences. Studies performed on Y chromosome heterochromatin are promising. They have allowed us to identify the existence of highly repetitive DNA sequences, characteristic only for this chromosome. By reassociating the DNA of a man with an excess of the DNA of a woman, Kunkel et al. [1976, 1977] showed that male DNA sequences, remaining single-stranded, can reassociate with male DNA, but not with female DNA. Experiments conducted with isolated DNA from cells containing the aberrant Y chromosome, allowed the localization of the "male" DNA sequences on the long arm of this chromosome. Using a different method, Cooke [1976] isolated this type of DNA with restriction enzymes. Using electrophoresis, after the application of the enzyme Hae III, the author identified two segments, being DNA fragments corresponding to sequences 3.4 × 103 and 2.1 × 103, and consisting of approximately 3,000 nucleotide pairs (copies). These segments are not found in DNA of female origin. Through appropriate purification, two specific DNA fractions of the Y chromosome, called Fr Y1 DNA and Fr Y2 DNA, were obtained. McKay et al. [1978], studying individuals who had Y chromosomes of various sizes, showed that these fractions of DNA are localized in long arm heterochromatin and that the amount of DNA corresponds to the length of the heterochromatin segment of the Y chromosome.

Mitchel and Bostock [1977] analyzed satellite III using the restriction enzyme Hae III and found that it consists of monomers 200 nucleotide pairs in length. Satellite DNA III in normal men contains two large fractions that are absent in DNA in normal women. Bostock et al. [I978] showed that the male-specific DNA sequence has a size of 3,500 nucleotide pairs; it repeats about 1,000 times on the Y chromosome and makes up almost I5 % of the mass of the Y chromosome. Experiments with aberrant Y chromosomes have shown that Y-specific DNA is probably localized in the weakly fluorescent region (Y q12) of the long arm of the Y chromosome, which is closely adjacent to the brightly fluorescent region.

The above information regarding the chemical characteristics of the DNA of HRs and their localization in the human karyotype was obtained primarily using the *in situ* hybridization method of nucleic acids [Gall and Pardue 1969; John et al. 1969]. Some additional information was obtained through radioautography, fluorescence, and immunochemistry. The theoretical basis of the first method is the idea that labeled H3-thymidine and H3-deoxycytidine are selectively included in parts of chromosomes whose DNA is rich in either A-T or G-C pairs of nitrogenous bases.

The fluorochrome method, as a tool for analyzing the composition of the nitrogenous bases of DNA in the chromosome, begins with the work of Caspersson and his collaborators [1968] who used quinacrine-mustard in the hope of identifying sites enriched with guanine. Furthermore, this method, based on different degrees of fluorescence depending on the composition of the nitrogenous bases of the DNA, has included the use of acridine [Weisblum and Haseth 1972; Selander 1973], benzimidazole derivatives [Weisblum 1974], and proflavine [Disteche and Bontemps I974].

In model experiments with pure DNA preparations enriched with corresponding pairs of nitrogenous bases, these fluorochromes showed that DNA containing A-T but not G-C base pairs actively fluoresces. However, later experiments by the authors, who proposed certain fluorochromes for studying the composition of nitrogenous bases in DNA, as well as other researchers, showed that the extrapolation of model experiments using pure DNA preparations to the whole chromosome was problematic [Selander and de la Chapelle 1973; Weisblum 1974; Disteche and Bontemps 1974].

Dev et al. [I972] proposed an immunochemical method, based on the use of an antiserum specific to one nitrogenous base or another in DNA. However, the reliability of this method is doubtful. According to data provided by two research groups [Dev et al. 1972; Schreck et al. 1973], pericentromeric chromosome regions, especially those rich in satellite DNA, do not respond to the corresponding antibodies. In general, it can be said that the data obtained by these last three additional methods have, at best, confirmed the data from the *in situ* hybridization method.

2.5. Human chromosomal C-heterochromatin regions (C-HRs)

The starting point for the development of the C-staining technique was work on *in situ* hybridization of nucleic acids on chromosome preparations [Pardue and Gall 1970]. In the process of *in situ* hybridization, the DNA of cell nuclei or metaphase chromosomes acts as a kind of immobilized receptor for reaction with complementary radiolabeled RNA or DNA. The position of the label after hybridization indicates those regions of the chromosomes that contain repetitive nucleotide sequences complementary to the introduced donor RNA. By conducting nucleic acid hybridization on the chromosomes of mice, Pardue and Gall [1970] discovered a very interesting morphological phenomenon. After the alkaline DNA "denaturation" of chromosomes necessary in such experiments and the subsequent thermal "renaturation" in standard salt solution (SSC), it turned out that only the pericentromeric regions of the mouse chromosome, which correspond to regions with "renatured" DNA, stain with Giemsa dye.

Arrighi and Hsu [1971] confirmed this observation on chromosomes of other animal and human species. The authors showed that, with such a procedure, structural heterochromatin, including centromeric heterochromatin, is stained. This method has been developed further in several works [Tanguay et al. 1971; Craig-Holmes and Shaw 1971; Sumner 1972). A modification of this method is the technique of specific staining of the pericentromeric region of human chromosome 9, which is located both in the metaphase and the interphase [Gagne and Laberge 1972; Bobrow et al. 1972].

Chromosomal C-heterochromatic regions (C-HRs) detected by the C-staining technique, in contrast to Q-heterochromatin, have, without exception, all chromosomes in the human karyotype [Paris Conference 1971, Suppl. 1975]. According to their localization, chromosomal C-HRs are divided into four groups: 1) C-heterochromatin of the secondary constrictions of the long arms of autosomes 1, 9, and 16; 2) pericentromeric C-heterochromatin (centromeric regions of all chromosomes without exception, including the Y chromosome); 3) C-HRs of short arms of all acrocentric chromosomes; and 4) C-heterochromatin segment of the distal part of the long arm of the Y chromosome (Fig. 2.5.1).

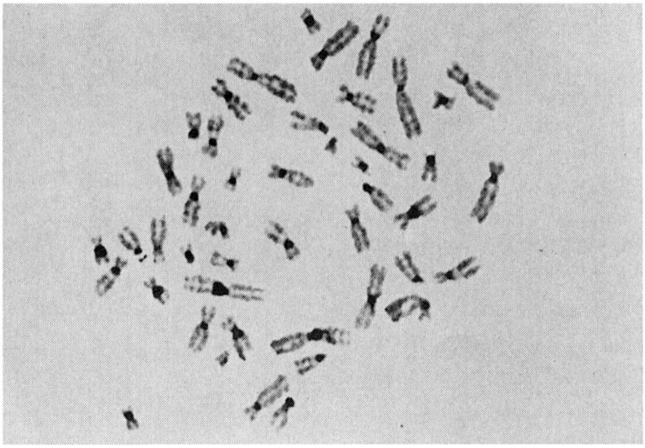

Fig. 2.5.1. Human chromosomal C-heterochromatic regions (C-HRs) after C-staining. C-HRs of chromosomes (dark-colored areas) are localized on all chromosomes in the human karyotype without exception.

As such, the variability of human chromosomal C-HRs consists of morphological variations of two types: a) a heteromorphism of C-band sizes and b) a heteromorphism of C-band localization relative to the arms or centromere regions of the corresponding chromosomes.

Heteromorphism in chromosomes 1, 9, and 16 by the size of their secondary constrictions was described in the early 1960s [de la Chapelle 1961; Ferguson-Smith et al. 1962]. Initially, these morphological features were also used to identify individual chromosomes (1, 9, and 16) [Patau 1960; de la Chapelle 1961; Muldal and Ockey 1961; Ferguson-Smith et al. 1962; Jacobs et al. 1967]. Some time later, the frequency of such chromosomal variants in populations became known, as well as the fact that their carriers do not have any noticeable phenotypic abnormalities.

After the development of the C-staining method, it became clear that these regions of chromosomes are large C-heterochromatic blocks with significant morphological variation, both in terms of size and localization (so-called "inversions") [Craig-Holmes et al. 1975; McKenzie and Lubs 1975; Buckton et al. 1976]. At the same time, the first attempts at their classification by size was made as small, medium, and large. However, the subjectivity of such an approach soon became apparent and the literature began recommending more reliable criteria for considering C-band sizes. As an acceptable semi-quantitative criterion in describing the size of C-bands, many researchers suggested using the sizes of some regions of individual chromosomes in the human karyotype: the long arm of chromosome 21 [Müller et al. 1975] or the short arm of chromosome 16 [Patil and Lubs 1977a]. Madan and Bobrow [1974] studied the polymorphism of sizes and variants of C-heterochromatin on the short arm of chromosome 9 (9p), using it as a standard. Metaxotou et al. [1978] took the same approach using the long arm of chromosome 21 (21q). Similar studies have been undertaken to study the polymorphism of the C-heterochromatin region of chromosome 1 [Aula and Saksela 1972].

Baliček et al. [1977, 1978] and Podugolnikova et al. [1979] attempted a more objective assessment of the size variability of C-bands of chromosomes 1, 9, and 16 by direct quantitative measurement. This allowed the authors to establish a certain correlation between the degree of contraction of chromosome arms and the size of C-heterochromatin segments. They concluded that the sizes of C-heterochromatin segments in populations obey the Gauss distribution and that, by their nature, the sizes of C-HRs variants are non-discrete. The continuity of the nature of the distribution of the length of C-heterochromatic segments has been discussed a number of times [Buckton et al. 1976; Lelikova and Tsvetkova 1976; Jacobs 1977; Verma et al. 1978a, b; Podugolnikova et al. 1979].

A significant part of C-heterochromatin variants are localized in the centromere regions of chromosomes. However, the difficulty of identifying all pairs of chromosomes in a human karyotype after C-staining does not allow the registering of all cases of C-polymorphism. For this, it is necessary to preliminarily identify all chromosomes with the help of Q or G-staining and then to restain them using the C-technique [McKinzie and Lubs I975; Müller et al. 1975; Lubs et al. 1977]. In this regard, the only exception is the centromere region of the Y chromosome where there is the smallest C-HR in the human karyotype, the recording and registration of morphological variants of which are extremely difficult.

There is very little information about the variability of pericentromeric C-heterochromatin in human populations due to the difficulties of its registration under a light microscope. The Paris Conference [1971] and Arrighi and Hsu [1971] limited themselves to stating the fact that C-polymorphic bands are localized in the pericentromeric regions of each chromosome in the human karyotype. Nevertheless, information on the variability of localization of C-heterochromatin segments of some chromosomes is currently accumulating. Phillips [1980] and Herva [1981] provided quantitative data on the localization of C-polymorphic centromeric heterochromatin on all human chromosomes. More detailed information on C-HRs of human chromosomes is given in reviews by Jacobs [1977], Baliček et al. [1978], Erdtmann [1982], Ibraimov and Mirrakhimov [1985], Prokofyeva-Belgovskaya [1986], Verma [1988], and Bhasin [2005].

Information on the chromosomal C-HRs in the human genome is presented here only to inform the reader about the existence of genetic polymorphism in this type of constitutive heterochromatin in the human population. However, this issue will not be considered further, since, due to the difficulty of identifying all chromosomes in a human karyotype after C-staining, it is not possible to objectively assess the existence of interindividual and interpopulation differences in the quantitative content of chromosomal C-HRs. In addition, it turns out that human populations do not differ from each other in the total number of C-HRs in the genome [Ibraimov and Mirrakhimov 1982a; Erdtmann 1982; Cavalli et al. 1985).

2.6. The phenomenon of chromosomal differential fluorescence

Studies seeking to elucidate the mechanism of selective inclusion of quinacrine mustard in certain regions of chromosomes have shown that Caspersson's initial idea of the nature of differential fluorescence [Caspersson et al. 1968] is wrong. First, intense differential fluorescence of the chromosomes, completely identical to that given by quinacrine mustard, can be obtained using simple quinacrine [Pearson et al. 1970; O'Riordan et al. 1971; Sumner et al. 1971]. Second, a series of studies performed *in vitro*, showed that DNA fractions rich in A-T, and not G-C, with pairs of nitrogenous bases, possess the ability to fluorescence intensely [Weisblum and de Haseth 1972]. Finally, it was found that chromosome regions intensively fluoresce in *Samoaia leonensis* (Drosophilidae), including a radioactive label at the end of the synthesis period (i.e. during replication of chromosomal heterochromatin regions) using H3-thymidine. During the replication period, these parts of chromosomes remain unchanged if using H3-cytidine [Ellison and Barr 1972].

Despite a large amount of research, up to now the nature of the differential fluorescence of chromosomes remains unclear. However, there have been a number of important advances in our knowledge in this area. First of all, it has been shown that DNA enrichment in heterochromatic regions of chromosomes by A-T pairs of nitrogenous bases is a necessary condition, although not sufficient for the occurrence of intense fluorescence when processing derivatives of acridine, whereas the enrichment of G-C pairs can be sufficient for the absence of such fluorescence. It has also been shown that the density of fluorescence is strongly influenced by the condensation density of the heterochromatic regions of the chromosomes themselves [Holmquist and Dancis 1979].

At the same time, it has been shown that not all regions of chromosomes characterized by a high content of A-T base pairs show intense fluorescence after Q-staining. In this respect, Hoechst 33 258 fluorochrome, which invariably binds to A-T enriched pairs of chromosomes, is more specific. When staining with quinacrine, large heterochromatin blocks of chromosomes 1, 9, and 16 enriched by A-T pairs are not detected [Chen 1977]. It turns out that the degree of fluorescence intensity during Q-staining depends not only on the presence of A-T pairs, but also on the nature of their alternation with G-C pairs and spatial localization [Selander and de la Chapellé 1973; Weisblum 1973].

Even before acridine dyes were used for the differential staining of chromosomes, *in vitro* experiments showed that the intercalation of the mustard group into the DNA molecule requires the latter to exist in the form of a double chain [Weisblum and de Haseth 1972], although there are reports of the interaction of acridine with partially denatured DNA [Waring 1965].

Regarding the dependence of acridine fluorescence on other components of chromosomes, it is now generally accepted that they are not decisive in the formation of Q-patterns. It has been determined that the intensity of Q-fluorescence does not change when chromosomes are treated with RNase, i.e. there is no dependence on the presence of RNA [Comings 1971]. The literature indicates that the protein components of chromosomes have some influence on the intensity of fluorescence during Q-staining. The number of sites available for intercalation of acridine heterocycles is lower for deoxyriboproteins than for DNA. Non-histone proteins of the DNP-complex change the fluorescence intensity during Q-staining [Gottesfield et al. 1974], but it has been shown that HRs enriched with satellite DNA do not contain non-histone proteins. As such, it should be concluded that the features of the protein components of chromosomes do not have a significant effect on the identification of Q-positive segments.

In this regard, it is necessary to dwell on one more aspect of the differential fluorescence of chromosomes. With the discovery of chromosomal differential staining by acridine derivatives, the question arose: do all heterochromatic regions of chromosomes have the property of intense fluorescence? Even the first observations of the chromosomes of *Drosophila melanogaster*, a number of plants, and in humans, in which the exact localization of the heterochromatic regions had been previously established by other methods, showed that not all HRs of chromosomes are capable of intense fluorescence. For example, it was clearly demonstrated on polytene chromosomes of Diptera (Drosophilidae) that only some heterochromatic discs have the property of intense fluorescence [Adkisson et al. 1971]. Similar results were obtained for plant chromosomes [Vosa 1971].

No less convincing data was obtained for the human karyotype. Thus, with the help of C-staining, it was established that constitutive heterochromatin is present in the pericentromeric regions of all human chromosomes. It has been shown that individual human chromosomes have very large variable HRs (the proximal part of the long arms of chromosomes 1, 9, and 16, and the distal part of the Y chromosome) [Arrighi and Hsu 1971]. With Q-staining of human chromosomes, with the exception of the distal part of the long arm of the Y chromosome, these large heterochromatic regions do not fully fluoresce [Paris Conference 1971]. As such, clear evidence has been obtained that, despite the well-known tendency of many chromosomal HRs to conjugate with each other in interphase nuclei, they display subtle differences in the biochemical composition or physical state.

2.7. The technique of chromosomal Q-staining

To date, about 25 compounds have been tested as agents capable of producing differential fluorescence in human chromosomes [Miller et al. 1973]. It has now been established that, in the main, only derivatives of the acridine dye, such as quinacrine dichloride ("Atebrin") and quinacrine mustard, have the property of giving the most distinct, bright, and reproducible Q-staining of chromosomes. Quinacrine, a well-studied acridine derivative, was, until recently, used as an antimalarial drug. The antimalarial activity of quinacrine is based on its ability to bind to DNA and block RNA synthesis [Kurnick and Radeliffe 1962; O'Brien et al. 1966]. Quinacrine mustard binds more strongly to native DNA than to depolymerized DNA [Morthland et al. 1954]. This property of quinacrine and quinacrine mustard is preserved in the pH range 3.7-7.4, that is, it barely depends on the pH of the medium [Peacocke and Scarret 1956]. It is possible that all derivatives of acridine dye induce an identical picture of differential fluorescence in human chromosomes [Miller et al. 1971]. Today, quinacrine mustard is primarily used as the fluorochrome. Caspersson and his colleagues [Caspersson et al. 1969a, b] showed that quinacrine mustard gives the most clear, stable, and reproducible differential fluorescence, since it is more resistant to the action of ultraviolet rays than quinacrine. Therefore, up to now, the Q-staining technique developed by the Caspersson group is still being used with almost no significant changes (Fig. 2.7.1).

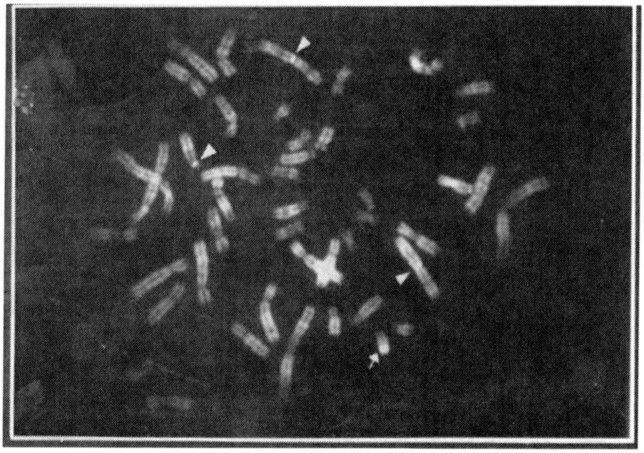

Fig. 2.7.1. Chromosomal Q-heterochromatin regions (Q-HRs) after Q-staining. This microphotomicrograph shows medium-sized Q-HRs and the 5th degree of fluorescence intensity on the long arms of chromosome 3, the centromeric region of chromosome 13, and on the long arm of chromosome Y (indicated by arrows).

The Q-staining technique itself is very simple. Here we present Caspersson's original method [Caspersson et al. 1970]. The chromosome preparations, after processing through a series of alcohols (starting from absolute alcohol) and buffer (phosphate buffer McIlwein/citrate buffer, pH 7.0), are immersed in a dye solution. The stock aqueous solution of quinacrine mustard is mixed with a buffer at the correct ratio to obtain a final dye concentration of 50 μg ml and then used to stain for 20 minutes at 20 °C. The slide is then rinsed three times in the buffer solution and coated with a thin coverslip. The differential fluorescence of the chromosomes is observed under a fluorescence microscope.

This technique is considerably simplified and almost standardized and the slide (chromosome preparation) is immersed for 5 minutes in a 0.005 % aqueous solution of quinacrine mustard. Staining can be obtained simply by applying a few drops of dye directly onto the surface of the slide. If the dye is made by immersing a slide into the dye solution, the slide is rinsed three times in distilled water or in the buffer (25 seconds each). It is preferable to use a buffer solution with a pH of 6.0, since, when the slide is rinsed with distilled water, the chromosomes often "swell". The slide, with remnants of water or buffer solution (without drying or washing), is covered with a cover glass so that no air bubbles remain under it. Excess fluid is removed with filter paper. The slide is then immediately placed under the lens of the microscope to prevent the liquid contained between the slide and the cover glass from drying out.

To undertake Q-staining, we can use the compound propylquinacrine mustard (PQM), which is also an acridine derivative. Mikelsaar [1981] has argued that PQM-stained cells for radiation exposure are five times greater than when stained with quinacrine (Q) and quinacrine mustard (QM).

2.8. Human chromosomal Q-heterochromatin regions (Q-HRs) and their identification

A remarkable peculiarity of the human genome (*Homo s. sapiens*) and the genomes of the two higher primates—chimpanzees (*Pan troglodytes*) and gorillas (*Gorilla gorilla*)—is that their karyotypes have areas in the chromosomes where Q-heterochromatic regions can be found. The techniques of Q and C-staining of chromosomes, apparently, reveal different types of constitutive heterochromatin, if only because C-heterochromatin is found in the genome of all higher eukaryotes, whereas Q-heterochromatin has a very limited distribution.

Pearson et al. [1971] investigated 27 species of mammals for the presence of Q-heterochromatic regions with the help of quinacrine. Their most surprising finding was that only humans, chimpanzees, and gorillas have autosomes with brightly fluorescent segments and only humans and gorillas have an intensely fluorescent Y chromosome. A chimpanzee has a small Y chromosome that does not produce intense fluorescence after Q-staining. In humans, such brightly fluorescent segments are located only on seven pairs of autosomes and the Y chromosome [Paris Conference

1971; Pearson 1973]. Subsequent studies have fully confirmed these observations.

C-heterochromatin is always present on all human chromosomes, varying only in size and, very rarely, in localization. Q-HRs can be completely absent on any of the Q-polymorphic chromosomes without any noticeable pathological or other phenotypic consequence for the carrier, whereas the complete absence of C-HR, even on one chromosome, is an extremely rare phenomenon [Paris Conference 1971, Suppl. 1975].

Corresponding studies have shown that Q-HRs and C-HRs are not always completely identical in localization and size on those chromosomes where they are located on the same segments (the Y chromosome, autosome 3, and the short arms of acrocentric chromosomes) [Verma and Dosik 1982].

C-HRs are known to constitute about 20 % of the human genome [John 1988]. According to some authors, chromosomal Q-HRs of a certain class may be completely absent in a significant part of individuals in human populations [Yamada and Hasegawa 1978; Ibraimov et al. 1982, 1986]. As such, there are sufficient grounds to assert that chromosomal Q-HRs and C-HRs do have significant qualitative and quantitative differences. Below is a detailed description of the morphology of human chromosomal Q-HRs.

In the human karyotype, only seven pairs of autosomes and the Y chromosome have Q-heterochromatic regions. For almost 60 years, since the publication of the work of the Caspersson group, it seems that there has not been a single day that any cytogeneticist (and not just cytogeneticists) anywhere in the world, for any one of a number of reasons, has observed Q-stained human chromosomes under a fluorescence microscope. However, during this time, no one has been able to see Q-HRs in other areas of the normal human karyotype, except in 12 potentially polymorphic loci of seven autosomes (3p11q11, 4p11q11, 13p11, 13p13, 14p11, 14p13, 15p11, 15p13, 21p11, 21p13, 22p11, and 22p13) and the q12 segment of the Y chromosome. In Fig. 2.8.1, potentially Q-polymorphic regions of chromosomes are indicated by arrows. Detailed information on the identification of non-polymorphic regions of homologous chromosomes in the human karyotype is given in the materials of the Paris Conference [Paris Conference 1971, Suppl. 1975; ISCN 1978].

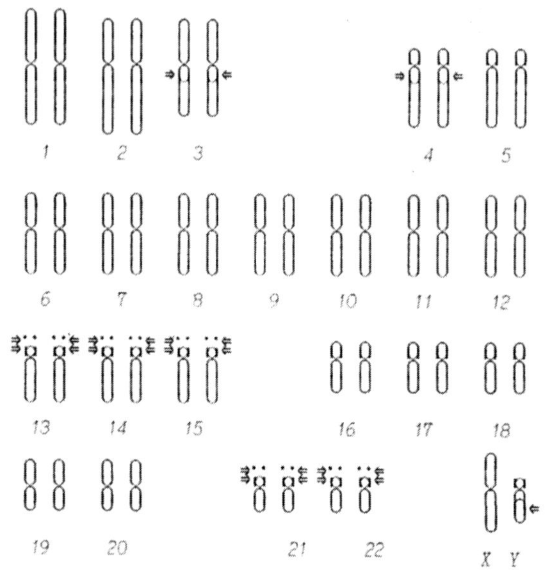

Fig. 2.8.1. Idiogram of the human karyotype. Using the Q-staining technique, Q-heterochromatic regions can be detected on 24 sites of seven pairs of autosomes (3, 4, 13, 14, 15, 21, and 22) and on the distal part of the long arm of the Y chromosome (these areas are indicated by arrows). Here is a male chromosome set. In women, instead of the Y chromosome, there is a second X chromosome, which does not have Q-heterochromatin.

The main morphological expression of the phenomenon of wide variability in human chromosomal Q-HRs is that individuals in populations differ in the number, location, size, and fluorescence intensity of Q-bands on 12 potentially polymorphic loci of seven pairs of autosomes and the q12 segment (distal part of the long arm) of the Y chromosome. It should be emphasized that, in human populations, there is no individual who would have Q-HRs simultaneously on all 25 potentially Q-polymorphic loci of the 46 chromosomes in his karyotype. Usually, the number of Q-variants in a human karyotype in a population ranges from 0 to 10 [Ibraimov and Mirrakhimov 1985].

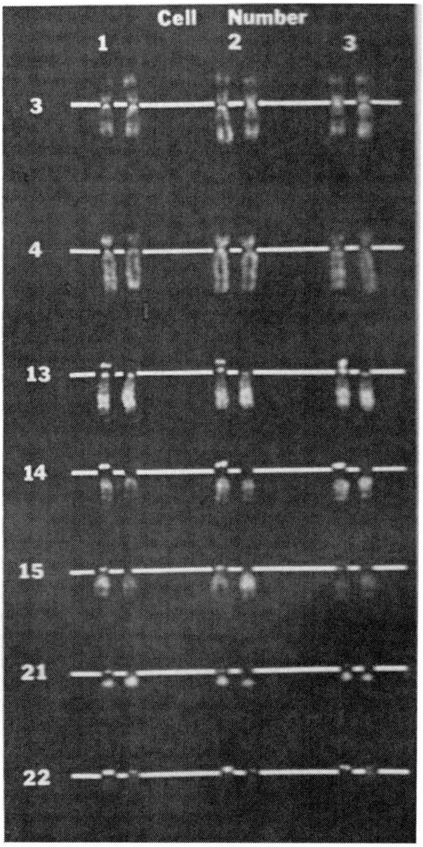

Fig. 2.8.2. Variable Q-heterochromatin regions of seven autosomes demonstrated by Q-staining (reproduced from Verma 1988 with permission of the publisher).

The best studied cytological characteristic of chromosomal Q-HRs is the features of localization in the karyotype. As has already been emphasized above, Q-HRs are not available on all chromosomes of the human karyotype. If we follow the conventional numbering of homologues, then the first Q-polymorphic autosome is chromosome 3. Usually, Q-variants are found on the centromeric region of its long arm (3p11q11 segment). However, in some populations, individuals have been found in whom the Q-heterochromatin band is localized on the pericentromeric region of the short arm of this chromosome [Allderdice 1973; Soudek and Sroka 1978; Ibraimov et al. 1982, 1986]. No other form of localization of Q-HRs on chromosome 3 has been detected [ISCN 1978; Ibraimov and Mirrakhimov 1985].

On chromosome 4, Q-HRs are mainly found in the pericentromeric region (segment 4p11q11). Although, as can be seen in the above idiogram (Fig. 2.8.1), a Q-HR is indicated on the same segment as on chromosome 3, in real populations Q-variants of chromosome 4 have three forms of localization: on the centromeric region; on the proximal part of the short arms; and on the proximal part of the long arms [Bardham et al. 1981; Ibraimov and Mirrakhimov 1985].

On acrocentric autosomes (13-15, 21, and 22), Q-HRs are localized mainly in the centromeric regions of the short arms (p11) and satellite regions (p13). However, Q-variants can often be found in the centromeric region of chromosome 22 [Al-Nassar et al. 1981]. No other localization of Q-HRs has been found on normal human chromosomes.

Chromosomal Q-HRs are very variable in terms of fluorescence intensity and size of Q-variants. At one time, the Paris Conference on Standardization in Human Cytogenetics recommended distinguishing Q-variants of chromosomes according to these characteristics [Paris Conference 1971]. This document offered a five-degree system for the classification of Q-variants (Table 2.8.1).

Table 2.8.1. The five-degree system for evaluating the variability of the size and fluorescence intensity of human chromosomal Q-HR variants.

Intensity level	(Paris Conference 1971)
1	Negative (no or almost no fluorescence)
2	Pale (as on distal 1p)
3	Medium (as the two broad bands on 9q)
4	Intense (as the distal half of 13q)
5	Brilliant (as on distal Yq)

As can be seen in this table, there are fairly clear criteria for identifying Q-HRs by the intensity of their fluorescence. However, this document does not give any specific recommendations for taking into account the size of Q-variants and does not report anything about the variability of chromosomal Q-HRs according to the peculiarities of their localization (see Section 2.16).

Soon after the introduction of the Q-staining technique in human cytogenetics, it became clear that chromosomal Q-HRs differ in size, both in the karyotype of a single individual and in the karyotype of different individuals in human populations. As far back as 1971, a five-degree system for the classification of the size variability of the chromosomal Q-HRs was proposed [Paris Conference 1971]. The recommendations of the Paris Conference on this issue were further developed in a number of papers [Müller and Klinger 1976; Yamada and Hasegawa 1978]. Over time, it became obvious that it was almost impossible to strictly follow the five-degree system of the Paris Conference because of chromosomal Q-HR variability. Therefore, for comparative population cytogenetic studies, chromosomal Q-HRs with only 4 and 5 degrees of fluorescence intensity have been taken into account, since their calculation and registration do not cause any great methodological difficulty [Schnedl 1971; McKenzie and Lubs 1975; Müller and Klinger 1976; Buckton et al. 1976; Lubs et al. 1977]. Yamada and Hasegawa [1978] were, perhaps, the first to recognize that there are individuals in human populations with different numbers of Q-HRs in their karyotypes. According to their observations, the number of Q-HRs in a human karyotype can range from 0 to 8—this has since been fully confirmed [Al-Nassar et al. 1981; Ibraimov et al. 1982, 1986] (for more information, see Section 2.14).

After Q-staining 1/2 to 2/3 of the long arm (q12 segment) of the Y chromosome gives off bright fluorescence. The variability of the Q-HR of the Y chromosome is mainly due to the variability in size of the brightly fluorescent q12 segment of the long arm, since the fluorescence intensity of this Q-heterochromatic segment always corresponds to the 5th degree [Paris Conference 1971] (for more details, see Section 2.16).

2.9. Calculation and registration of chromosomal Q-HRs

The calculation and registration of chromosomal Q-HRs is a very complex and not yet completely solved methodological problem, which all researchers in the field, without exception, face. Most of the difficulties in studying the variability of the morphology of human chromosomal Q-HRs are found at this stage of study. The calculation and registration of Q-variants starts from the moment of Q-staining of chromosomes. We can assume that some of the difficulties associated with the Q-staining technique have mostly been overcome or are surmountable. This all depends on the quality of the chromosomal preparation, the fluorescence microscope, and the fluorochrome used.

Chromosomal Q-HR variability can be directly observed under the microscope and, more recently, on a computer screen. There is still no consensus about which of these approaches gives an advantage. Starting with the pioneering research of the Caspersson group, many authors of population studies have analyzed Q-HR variants on photographic prints [McKenzie and Lubs 1975; Lin et al. 1976; Verma and Dosik 1976; Lubs et al. 1977]. However, most cytogeneticists prefer to undertake a visual analysis of Q-differentiated chromosomes directly under the microscope [Geraedts and Pearson 1974; Mikelsaar et al. 1974, 1975, 1977; Müller and Klinger 1975; Buckton et al. 1976; Schwinger and Wehner 1976; van Dyke et al. 1977; Yamada and Hasegawa 1978; Ibraimov et al. 1982].

In our opinion, if we are talking about comparative population cytogenetic studies, preference should be given to a direct visual analysis of Q-variants. There are several reasons for this. First, very often small Q-variants cannot be detected on photographic prints; it is also difficult to identify the actual ratio of fluorescence intensity levels between Q-heterochromatin. McKenzie and Lubs [1975] reported that only the 3p11q11 segment can achieve the 5th degree of fluorescence intensity and the degree of fluorescence of the remaining Q-HRs of the autosomes does not exceed the 4th degree (i.e., the fluorescence intensity of the distal part (13q) of the long arm of chromosome 13). This is an obvious effect associated with the analysis of Q-variants in microphotographs, since, with direct visual analysis under a microscope, it is not difficult to make sure that all 12 polymorphic loci of the seven autosomes contain Q-HRs at the 5th degree of fluorescence intensity (i.e., the same as on the distal part of the (q12) Y chromosome). Secondly, the analysis of Q-variants in micrographs, at least in the first years of the application of Q-staining (when "Atebrin" was used as a fluorochrome) was justified due to the rapid fading of Q-stained bands under the action of ultraviolet rays. This severely limited possibilities for visual analysis. Finally, for comparative population studies, the method of visual analysis of chromosomal Q-HRs is preferable because of its speed and greater accuracy in identifying Q-HR variants [Geraedts and Pearson 1974; Buckton et al. 1976; van Dyke et al. 1977; Yamada and Hasegawa 1978; Ibraimov and Mirrakhimov 1985].

The next question relates to the selection of Q-stained metaphase chromosomes for analysis. Usually, metaphase plates with moderately spiralized chromosomes are selected, with optimal spread and a clear picture of differentiation for visual analysis. Finding such metaphase plates with good chromosomal preparations is not difficult, but, at the same time, prometaphase and, in particular, too-contracted metaphase chromosomes should be avoided. In this latter case, there is a danger of "losing" part of the Q-variants if they are small in size or their fluorescence intensities are at the 4th degree.

Furthermore, it is necessary to solve the following important questions: how many metaphase plates need to be analyzed to determine the exact number of chromosomal Q-HRs, the intensity of their fluorescence, and their localization in a given individual? The fact is that there is an intercellular variability in the manifestation of Q+ bands, the reasons for which are not entirely clear. This means that some Q-HRs variants in a given individual appear only in a part of the cells. The only thing that is known is that such intercellular variability is not the result of mosaicism of the studied tissue. Here, it should be emphasized that intercellular variability occurs infrequently and usually affects Q-HR variants of very small sizes. It is especially noticeable when, for one reason or another, there are mainly contracted or poorly differentiated metaphase chromosomes in the preparation. This circumstance, in one way or another, is confronted by all researchers of comparative population studies. Therefore, the first researchers of the variability of chromosomal Q-HRs in human populations decided that, before ascertaining the causes of intercellular variation, it was necessary to investigate at least 10 cells per individual. As such, the carrier of this type of Q-HR variant can be considered only for an individual in whom this variant is in no less than half of the studied cells.

This question is still far from receiving a definitive answer. So far, the number of cells required for analysis is determined by each author almost arbitrarily, since it is clear that the number 10, adopted by some researchers, does not include any objective pattern related to the biological nature of the wide hereditary variability of chromosomal Q-HRs in human populations. We, for example, have been convinced by our own experience that in order to obtain reproducible results, it is necessary to analyze at least 20 metaphase plates [Ibraimov et al. 1982; Ibraimov and Mirrakhimov 1985]. It is possible that not just 20, but perhaps even 10 cells, are more than is necessary for comparative population cytogenetic studies. On the other hand, to study the inheritance of Q-variants in families, mosaicism, or in some complex applied research, such as in the case of the definition of controversial paternity, even 20 metaphases may not be enough. All this must be remembered when planning and discussing the results of any study. It is even better if these methodological details are appropriately specified by the authors themselves in their reports and publications.

As noted above, chromosomal Q-HR variants can have different degrees of fluorescence intensity. At one time, the Paris Conference on Standardization in Human Cytogenetics recommended registering all types of Q-variants, from the first to the fifth degrees of fluorescence intensity (see Table 2.8.1). The first authors of population cytogenetic studies attempted to adhere to the recommendations of the Paris Conference. However, it soon became apparent that the error-free identification of Q-variants of the first three degrees of fluorescence intensity was practically impossible, especially when analyzing micrographs of chromosomes. The experience of most cytogeneticists has convinced us that Q-variants with intense and bright fluorescence are amenable to more or less accurate identification. Currently, all researchers accept Q-variants with only 4 and 5 degrees of fluorescence intensity for chromosomal Q-heterochromatic regions [Paris Conference 1971, Suppl. 1975; Ibraimov and Mirrakhimov 1985].

Unfortunately, this does not resolve all the difficulties associated with the registration of human chromosomal Q-variants; here, we are dealing with two more. Chromosomal Q-HRs variants are analyzed immediately during the first hour after staining, that is, before the liquid enclosed between the slide and the cover glass dries out. Of course, if desired the slide can be re-stained. However, such a procedure, as a rule, impairs the quality of chromosomal Q-differentiation.

Many researchers have paid attention to the fact that optimal Q-staining is obtained with a relatively short preparation time. Our own experience has shown that satisfactory Q-differentiation of chromosomes can be obtained using preparations of not more than two months old, if they have been stored at room temperature. Analysis of slides with different shelf lives, obtained from the same individuals, has convinced us that during long-term storage (more than half a year), the degree of fluorescence intensity of Q-heterochromatin bands changes and, therefore, their initial number may decrease. This applies mainly to Q-bands of small sizes.

There is one more important circumstance connected to obtaining optimal Q-staining for the identification of Q-HRs. Obtaining the clear longitudinal differentiation of chromosomes is often hampered by the strong background fluorescence of the glass surfaces, which is associated with cell debris destroyed during the processing of a whole blood culture. Some cytogeneticists avoid studying the polymorphism of chromosomal Q-HRs precisely because of this methodological problem. Our many years of experience have shown that when working with whole peripheral blood cultures, it is useful to removal red blood cells before the cell fixation stage [Ibraimov 1983]. This technique significantly improves the quality of the differential fluorescence of chromosomes, as evidenced by the results of recent studies on the behavior of cells during their processing to obtain preparations of chromosomes [Ami et al. 2014].

2.10. Terminology

ISCN [1978] recommends that human chromosomes having heterochromatic regions varying between homologues in populations are called variable or heteromorphic chromosomes. The recent literature more often talks of the polymorphism of chromosomal HRs [Prokofyeva-Belgovskaya 1986; Zakharov and Tsoneva 1982; Geraedts and Pearson 1974; Mikelsaar et al. 1974, 1975; Müller et al. 1975; Buckton et al. 1976; van Dyke et al. 1977; Ibraimov et al. 1982, 1986]. To describe the variability of Q-heterochromatic segments after Q-staining, the term "chromosomal Q-polymorphism" or simply "Q-variants" is usually used.

The term "heteromorphism", adopted by ISCN [1978], also implies the continuous nature of changes in fluorescence intensity and the size of chromosomal Q-HRs, which at one time were classified arbitrarily into five discrete levels (see Table 2.8.1). According to this international document, which adds to and continues on from two previous ones [Paris Conference 1971, Suppl. 1975], the terms "norm" and "heteromorphic" cannot be opposed to each other. Each chromosome exists in a population in many variants. Therefore, the terms "norm", "heteromorphic", or simply "variant", with respect to human chromosomes, mean the form of its existence in the broad sense of the word. In the narrow sense, the term "variant" is applied to a chromosome characterized by one of its variable signs, detected by one of the particular methods of cytogenetic analysis—by routine or differential staining, autoradiography, or *in situ* hybridization of nucleic acids, etc. With respect to Q-staining, the "variant" is often described as a chromosomal band with bright and intense fluorescence, implying that the non-fluorescent segments represent the "norm". ISCN [1978] recommends calling the Q-staining technique using quinacrine, quinacrine mustard, or propylquinacrine mustard as a fluorochrome, QFQ-staining (Q-banding by fluorescence using quinacrine mustard).

We mainly use the term "variability" or "polymorphism", instead of the term "heteromorphism" recommended by ISCN [1978] and as used in some publications. In our opinion, the term "heteromorphism" is a somewhat limited concept and carries mostly a morphological meaning, whereas the term "polymorphism" is more suitable for describing monogenic tissue and cellular isoantigens, electrophoretic variants of proteins, enzymes and DNA. However, we have deliberately allowed ourselves some arbitrariness in the use of the terms "Q-polymorphism", "Q-variants", "Q-heterochromatin", "Q-heterochromatin regions", "Q-polymorphic variants", "Q-bands", and "Q-heterochromatin segments", which are synonymous, since these terms are taken from those literary sources that are analyzed in this work. We believe that such “freedom” with cytogenetic terminology will remain justified until the biological role of the wide hereditary variability of chromosomal Q-heterochromatin regions (Q-HRs) in human populations is clarified, at least in general terms.

In this book, the term Q-heterochromatin or Q-HRs refers to chromosome regions that display only intense (4th degree) and bright (5th degree) (“brilliant”) fluorescence. Areas of chromosomes with fluorescence intensity of degree 3 or below are considered to indicate the absence of Q-HRs in this Q-polymorphic locus of the chromosome under study.

2.11. Quantitative characteristics of chromosomal Q-HRs used in comparative population studies

The detection of the wide variability of chromosomal Q-HRs in human populations with the help of Q-staining, first of all, raises the question of how to create a rational and unified system for the quantitative accounting of Q-variants necessary for statistical analysis. Comparison of the frequencies of different types of polymorphic variants of chromosomal Q-HRs in different groups of individuals and populations (norm and pathological; sex, age, racial/ethnic groups, etc.) is important for studying the biological role of the polymorphism of HRs in the human genome. It goes without saying that any comparative cytogenetic study is unthinkable without comparable quantitative data. During the first four decades of the study of chromosomal Q-HR variants, a system of their accounting spontaneously took shape and a number of quantitative characteristics proved to be generally accepted, while other characteristics were considered only in individual works. The system of quantitative characteristics of polymorphic variants of human chromosomal Q-HRs considered here relies primarily on the experience of our laboratory and, naturally, needs further improvement [Ibraimov and Mirrakhimov 1985; Ibraimov et al. 1990].

We highlight the following quantitative characteristics of chromosomal Q-HR variants when studying human populations:

1. The frequency of Q-variants of chromosomes in populations. This is a common quantitative characteristic of the variability of chromosomal Q-HRs. It refers to population frequencies of Q-HRs at the 4th and 5th degrees of fluorescence intensity [Paris Conference 1971, Suppl. 1975] in each of the 12 potentially Q-polymorphic loci of the seven autosomes in a given sample. It is

also expressed as a percentage of the number of chromosomes analyzed separately for each Q-polymorphic locus.

2. The mean number of chromosomal Q-variants calculated per individual in a human population, is determined by dividing the total number of Q-variants found in this sample by the number of individuals examined;
3. The distribution of numbers of chromosomal Q-variants in populations. This characteristic simultaneously reflects the number of individuals with a different number of Q-variants in the karyotype and the scope of variability of chromosomal Q-HRs in populations, if by this we mean the minimum and maximum values of the numbers of Q-variants found in the corresponding samples [Ibraimov et al. 1986; 1990];
4. The frequency of inverted chromosome 3 in a population. In this case, we are talking about the frequency of the occurrence of the pericentric inversion of the Q-heterochromatin segment of chromosome 3, as first described by Allderdice [1973].
5. The size and intensity of Q-HRs in human populations based on the recommendations of the Japanese cytogenetic researchers Yamada and Hasegawa [1978] and the Paris Conference [Paris Conference 1971, Suppl. 1975].
6. The frequency of homo (+/+ and -/-) and heterozygotes (+/-) in human populations. Usually, homologues with Q-variants at the 4th and 5th degrees of fluorescence intensity are taken to be homozygous (+/+). Although, in such cases, Q-HRs are present on both homologues, they can nevertheless be completely different from each other in terms of the size and intensity of fluorescence. The concept of "heterozygote" is also arbitrary. In this situation, one of the homologues has Q-variants at the 4th and 5th degrees of fluorescence intensity, while the other does not. Where there is no such Q-variant on either homologue, then it is conventionally taken to be a homozygote (-/-) [Geraedts and Pearson 1974; Mikelsaar et al. 1974, 1978; Buckton et al. 1976].
7. The variability in size of the Q-heterochromatic segment (q12) of the long arm of the Y chromosome. In population studies, the variability of the Q-heterochromatin of a given chromosome is often studied separately from the Q-variants localized on the autosomes.

Differential data on the frequencies of Q-variants in individual loci were estimated using the hypergeometric criterion [Kendall and Stewart

1973]. Differences in the distribution of Q-variants by loci were estimated using the $\chi 2$ criterion when comparing two distributions. The comparison of samples to give mean numbers of Q-variants was made using Student's criterion (these issues are discussed in more detail in Ibraimov et al. 1990).

2.12. The frequency of chromosomal Q-HRs in human populations

To date, extensive data have been obtained on the frequency distribution of Q-variants across all seven Q-polymorphic autosomes in various human populations and covering all three racial groups [Schnedl 1971; Mikelsaar et al. 1974, 1975, 1977; Geraedts and Pearson 1974; McKenzie and Lubs 1975; Müller et al. 1975; Lin et al. 1976; Lubs et al. 1977; Verma and Dosik 1980b; Yamada and Hasegawa 1978; Tupitsina and Stobetsky 1980; Kruminia et al. Kroshkina, 1987; Al-Nassar et al. 1981; Herva 1981; Ibraimov and Mirrakhimov 1982 a, b, c; 1985; Ibraimov et al. 1982, 1986, 1990, 1991, 1997, 2013; Stanyon et al. 1988; Kalz et al. 2005; Decsey et al. 2006].

However, before considering the results of these studies, it is necessary to make a number of clarifications. Not all population studies were performed using a single cytogenetic methodology, which makes it difficult to conduct an objective comparison of their results. This situation is not explained by the lack of a unified international system for recording and registering chromosomal Q-HRs variants, since the first version of the recommendations of the Paris Conference appeared in 1971 and later, taking into account the accumulated information, this document underwent a number of changes [Paris Conference 1971, Suppl. 1975; ISCN 1978, 2005, 2009]. We have already devoted a special section to the detailed analysis of the reasons for this situation (see Sections 2.8 and 2.9).

Despite the currently known methodological difficulties found in the comparative analysis of data obtained in different laboratories, there are still no fundamental objections in the literature to the existence of some differences between human populations in the frequency distribution of Q-variants on the seven relevant autosomes. Table 2.12.1 presents data from the cytogenetic population studies of a number of authors who, judging by their publications, used similar methods for the registration and calculation of Q-variants, namely, they considered Q-bands at the 4th and 5th degrees of fluorescence intensity. In general, the results of all these studies lead to the conclusion that there is a non-random distribution of chromosomal Q-HRs among individual loci.

Table 2.12.1. The frequency of Q-HRs[a)] in 12 polymorphic loci of seven pairs of autosomes in different human populations.

Authors	Location of Q-HRs											
	3 cen	4 cen	13p11	13p13	14p11	14p13	15p11	15p13	21p11	21p13	22p11	22p13
Geraedts, Pearson, 1974 (n = 221)	48.4[a)]	2.7	50.0	-	14.3	-	21.5	-	24.4	-	21.9	-
McKenzie, Lubs, 1975 (n = 77)	75.9	40.9	44.2	2.6	4.6		1.3	-	2.6	-	7.1	1.9
Muller et al., 1975 (n = 261)	55.5	13.1	73.6	8.2	2.5	13.4	2.6	11.0	2.6	17.0	34.3	28.0
Buckton et al., 1976 (n = 950)	64.9	48.3	38.0	8.8	-	10.3	-	12.5	0.7	9.6	3.0	5.0
Lin et al., 1976 (n = 930)	55.8	14.1	31.4	1.8	0.8	0.2	0.2	0.9	0.1	1.1	0.3	0.3
Lubs et al., 1977												
(n = 205)	50.3	10.6	29.5	2.0	-	5.7	0.5	5.1	0.3	4.1	1.0	3.0
(n = 210)	58.2	4.6	50.3	5.1	-	6.7	1.5	5.1	1.0	6.7	2.0	7.3
Yamada, Hasegawa, 1978 (n = 400)	26.0	4.1	44.8	12.3	4.0	18.3	7.4	21.9	2.5	24.8	14.2	11.0
Al-Nassar et al., 1981 (n = 124)	79.0	41.1	71.0	11.3	2.4	10.5	1.6	8.1	11.3	1.6	4.8	8.9
Herva et al., 1981 (n = 91)	64.2	12.6	61.5	4.4	-	3.8	-	6.6	0.5	3.3	1.1	4.4
Tupitsina, Stobetsky, 1980 (n = 45)	43.3	-	33.3	14.4	14.4	31.1	5.5	25.5	22.2	35.5	3.3	15.5

Tsvetkova, 1981 (n = 55)	41.6	5.6	23.1	1.8	1.1	8.3	1.8	11.1	1.1	6.4	1.8	4.6
Kalz et al., 2002 (n = 100)* (n = 100)** (n = 67)***	 39.5 33.0 43.3	 6.5 12.5 10.5	 75.0 59.5 90.4	 - - -	 22.0 13.0 21.7	 - - -	 24.0 12.5 23.8	 - - -	 21.0 21.0 34.3	 - - -	 32.5 28.0 34.6	 - - -
Decsey et al., 2006 (n = 1131)	17.2	2.1	36.3	-	11.1	-	15.2	-	10.6	-	34.6	-

a) Frequency expressed as a percentage of the number of chromosomes analyzed.
* - Europoids; ** - Europoids; *** - Turks.

If we absolutize the differences in the data of individual research groups, then it seems impossible to detect any regularity when comparing their results. The resulting pattern, however, clearly reflects the existence of a certain order. There are loci in which the frequency of Q-variants significantly exceeds the frequencies characteristic of other loci. For example, none of the researchers in any of the surveyed populations found Q-HR frequencies in 3 cen and 13p11 loci lower than for any other loci, and, on the contrary, the lowest frequencies were usually consistently detected for 14p11, 21p11 loci (below 0.1).

This table also shows that, in all the studied populations, the overwhelming majority of Q-variants are localized in the chromosome pairs 3 and 13. Among the Japanese, the lowest frequency, compared to the European samples, of Q-variants on chromosome 3 (26.6 %) is observed. On the other hand, on the short arms and satellite regions of the acrocentric chromosomes in the Japanese population, Q-variants are more frequent than in Caucasians. The greatest heterogeneity of populations is observed with respect to the 4 cen and 22p13 chromosome segments. Obviously, some of these are likely to be associated with Q-variant identification errors—when re-examining the same sample by van Dyke et al. [1976], it was found that the greatest number of errors was in relation to chromosome 22. McKenzie and Lubs [1975] and Lubs et al. [1977] found an unexpectedly large discrepancy in the frequency of Q-variants of chromosome 4 with a twofold re-examination of the same population. On the other hand, judging by data from Yamada and Hasegawa [1978], it can be assumed that there are real differences between the Mongoloid and Caucasoid populations in the frequency distribution of Q-variants. However, the results of recent studies indicate that the picture of the distribution of frequencies of Q-variants in human populations is much more complicated than that presented in Table 2.12.1.

Usually, the differences detected for Q-HR frequencies in one or other locus are interpreted as reflecting actual differences between samples. For example, Lubs et al. [1977] studying samples for "white" and "black" Americans, noted statistically significant differences between them on chromosomes 3, 13, and 22; for all other loci, they noted a tendency to increased frequencies in the samples of black Americans. Buckton et al. [1976] noted a downward trend in the frequencies of Q-variants in all loci on a sample of individuals over 65 years of age when compared to newborns and adolescents. Such examples are numerous [Ibraimov and Mirrakhimov 1985].

As for the interpretation of these differences, there are two fundamentally different approaches. Proponents of the first assume that the nature of interpopulation heterogeneity with respect to the variability of chromosomal Q-HRs, is connected at the qualitative side of these differences, i.e., it depends on which specific locus sees heterogeneity [Buckton et al. 1976; Lubs et al. 1976; Nazarenko 1987]. The second approach is based on the idea of the absence of fundamental structural and functional differences between Q-HRs of different localizations in the karyotype and is expressed as follows: "first of all, the dose is important, not the localization of Q-HRs in one or another chromosome" [Ibraimov et al. 1982, 1986, 1990]. This issue is discussed in greater detail in Section 2.20.1.

2.13. The mean number of chromosomal Q-HRs in human populations

This quantitative characteristic of human chromosomal Q-HR variability, as with the previous one, is generally accepted in that the value of the mean number of Q-HRs, calculated per individual in a population, is given in many cytogenetic population studies.

However, the value of the mean number of Q-variants, calculated per individual in a population, as an important quantitative characteristic of chromosomal Q-HR variability, has not yet been properly evaluated [Ibraimov and Mirrakhimov 1985]. Almost all population studies are limited to presenting the numerical value of this quantitative characteristic without any discussion. A similar trend in data interpretation was noted in the works of Müller et al. [1975] and Nazarenko [1987]. Only Lubs et al. [1977], who found a significant increase in the mean number of Q-HRs among black Americans when compared to white Americans, attributed this tendency to a possible "founder effect".

Table 2.13.1.1 presents data on the mean number of Q-variants per individual of those authors who used similar criteria for recording and registering chromosomal Q-heterochromatin bands. On average, this value is 3.8. Trying to compare the data obtained by various authors, we inevitably face significant difficulties. The only conclusion that seems indisputable is that there are indeed differences between different human populations in the mean number of Q-variants per individual. Nevertheless, the numerical values of the Q-variants presented in this table should be treated with caution. First, as we have stressed repeatedly above, there is the incompatibility of quantitative data from different authors working in different laboratories (the possible reasons for such differences have been explained in Section 2.9). Secondly, the heterogeneous age, gender, racial, and ethnic composition of the samples should be noted (for more details, see Section 2.20).

Table 2.13.1. Mean number of chromosomal Q-HRs in different populations*.

Authors	Populations	Sample size	Mean number of Q-HRs	Maximum and minimum number of Q-HRs in individuals in the population
Geraeds, Pearson, 1974	Europoids, Holland	221	4.0	-
McKenzie, Lubs, 1975	Europoids (Newborns)	77	2.9	0 - 5
Buckton et al., 1976	Europoids Newborns 14 year olds 65 years and older	 482 109 210	 4.2 3.0 2.9	 - - -
Müller et al., 1975	Newborns, USA	357	5.2	-
Lin et al., 1976	Europoids (83 %), Newborns	230	2.1	-
Yamada, Hasegawa, 1978	Mongoloids, Japanese	400	3.0	0 - 8
Tupitsina, Stobetsky, 1981	Europoids, Uzbeks	45	4.0	-
Al-Nassar et al., 1981	Europoids, Arabs	146	5.0	1 - 9
Herva, 1987	Europoids, Finns	91	3.3	1 - 7
Nazarenko et al., 1987	Mongoloids, Khants	469	3.6	-
Kruminia et al., 1987	Europoids, Latvians	140	10.0	-
Own data	Mongoloids: Kyrgyz Chinese Kazakhs Mongols Chukchy Khakass	 1358 124 456 72 132 120	 2.5 3.5 3.6 3.4 2.2 2.5	 0 - 7 0 - 7 0 - 8 0 - 7 0 - 6 0 - 6

	Yakuts	127	1.9	0 - 5
	Khants	54	1.8	0 - 5
	Selkups	90	1.8	0 - 5
	Negroids:			
	Mozambiqueans	148	4.6	1 - 9
	Angolans	132	4.7	1 - 10
	Ethiopians	52	3.3	1 - 7
	Zimbabweans	34	4.7	1 - 7
	Guinea-Bissauis	13	4.6	2 - 7
	Russians	693	2.4	0 - 6

* Mean number calculated on the basis of the data provided by the authors of the study.

The question of the presence and degree of interpopulational differences in the mean number of chromosomal Q-HRs remains open. Samples surveyed by various authors are often heterogeneous in one important feature or another. This often negates the comparative value of even very thorough research. For example, after examining about 1,000 newborns, Lin et al. [1976] detailed all the characteristics of their sample, up to the age of motherhood, but those relating to ethnic composition is limited to indicating that 87 % of the surveyed individuals were Caucasian and the rest were Negroid or Mongoloid in origin. Al-Nassar et al. [1981] does not provide any information about the representativeness of three Arab communities (Kuwait) surveyed, commenting only that the samples were "random".

In this regard, studies in which such sources of uncertainty in the interpretation of the materials obtained are excluded, are of particular value; unfortunately the number of such studies is small [Buckton et al. 1976; Herva 1981; Ibraimov and Mirrakhimov 1985; Ibraimov et al. 1982, 1986, 1990, 1991; Stanyon et al. 1988]. In general, the characteristics of the samples can undoubtedly be considered most important in drawing any conclusions about the existence of differences between human populations in the total quantitative content of chromosomal Q-HRs and in the search for possible reasons for this phenomenon.

Finally, only in four papers do we find information about the distribution of Q-variants among the examined individuals; according to the magnitude of which one could judge the limits of variability of the quantitative content of Q-variants in the genome of this or that human population, which in itself would be of considerable interest (see Table 2.13.1.).

We surveyed more than two dozen native people living in different climatic and geographic conditions of Eurasia and Africa, and representing different racial, national, and ethnic groups. As can be seen in Fig. 2.13.1, the samples were significantly different from one another. Here, we once again note that the studied groups of individuals are comparable in age, homogeneous in ethnic composition within each sample, and, no less important, were examined under the strictest possible methodological conditions, according to uniform criteria for registration and calculation of Q-variants by the same cytogeneticist [Ibraimov 1993].

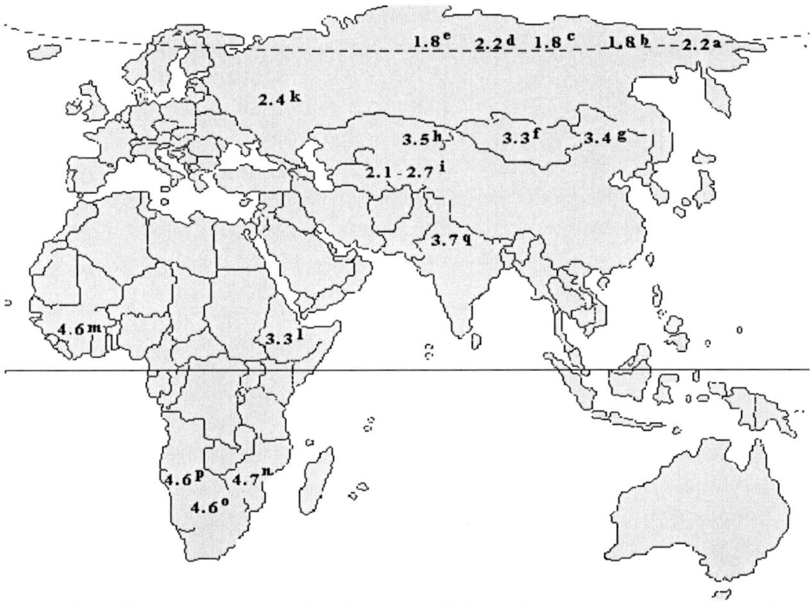

Fig. 2.13.1. The mean number of Q-HRs per individual in the native populations of Eurasia and Africa: a = Chukchi of Chukotsk (n = 132); b = Yakuts of the Yakut ASSR (n = 127); c = Selkups of Eastern Siberia (n = 90); d = Nenets of Eastern Siberia (n = 117); e = Khants of Eastern Siberia (n = 54); f = Mongolians of the MPR (n = 72); g = Chinese of northern China (n = 124); h = Kazakhs of southern Kazakhstan (n = 101); i = Kirghiz of the Pamirs and Tien Shan (n = 603); k = Russians of Bishkek (n = 200); l = Ethiopians of the Ethiopian Highlands (n = 52); m = Negroids of Guinea-Bissau (n = 13); n = Negroids of Mozambique (n = 148); o = Negroids of Zimbabwe (n = 34); p = Negroids of Angola (n = 132); q = Indians of northern India (n = 58).

First of all, we draw attention to the fact that the mean number of Q-HRs calculated per individual is statistically speaking significantly lower in populations living permanently at high geographical latitudes (Chukchi, Yakuts, Nenets, Selkups, and Khanty) and in high altitude areas (indigenous people of the Pamirs, Tien Shan, and the Ethiopian Highlands), compared to those living in the steppe zones of lowland Central Asia (Kazakhs of South Kazakhstan, Chinese of North China, and Mongols of the Mongolian People's Republic), North India and subequatorial Africa (Negroes of Mozambique, Angola, Zimbabwe, and Guinea-Bissau). Among Russians living in Kyrgyzstan, the value of this quantitative characteristic of chromosomal Q polymorphism is also reduced compared to the steppe populations of Central Asia. Among the indigenous inhabitants of the Ethiopian Highlands, this indicator is significantly less common than among the lowland Negroids of Africa, despite the fact that they live at similar geographical latitudes. Based on these data, as well as taking into account other quantitative characteristics of the Q-variants, we suggest the possible selective value of the amount of Q-heterochromatic chromosome material in the adaptation of human populations to certain extreme environmental factors, in particular to cold and hypobaric hypoxia [Ibraimov et al. 1982, 1986; Ibraimov and Mirrakhimov 1982 a, b, c, 1985] (for more details, see sections 2.20.2 and 3.3.3).

2.14. The distribution of numbers of chromosomal Q-variants in populations

It has been established that, in human populations, there are groups of individuals that show a complete absence or a high number of chromosomal Q-HRs [Yamada and Hasegawa 1978; Al-Nassar et al. 1981; Ibraimov et al. 1982, 1986, 1990]. In this case, it is a question of Q-HRs at the 4th and 5th degrees of fluorescence intensity [Paris Conference 1971]. As an illustration, we present a microphotograph of the chromosomes of four individuals, who differ from each other in the number of Q-HRs of chromosomes in the karyotype (Fig. 2.14.1).

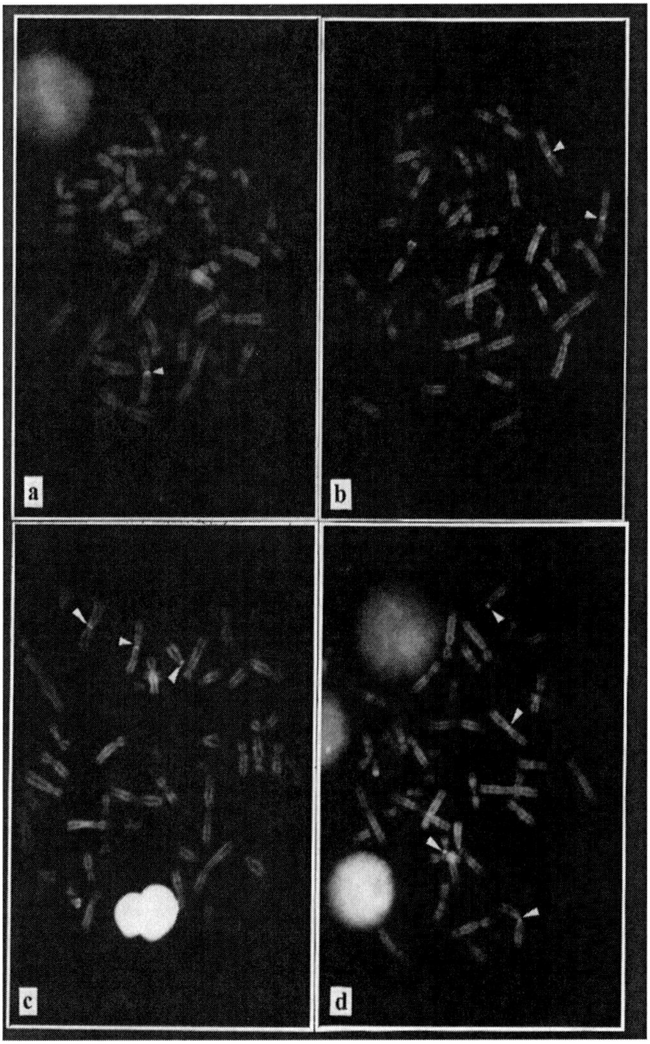

Fig. 2.14.1. Karyotypes of four human samples stained (Q-staining) with propylquinacrine mustard. (a) Metaphase of female (46, XX) having one chromosomal Q-HR. (b) Metaphase of human male (46, XY) having two chromosomal Q-HRs. (c) Metaphase of man (46, XY) having three chromosomal Q-HRs. (d) Metaphase of man (46, XY) having four chromosomal Q-HRs (reproduced from Ibraimov 2011 with permission of the publisher).

Unfortunately, not all population cytogenetic studies indicate the number of Q-variants in individual karyotypes (see Table 2.13.1). As can be seen in this table, only four papers contain information on the maximum and minimum number of Q-variants in the studied populations, but there is no data on the distribution of chromosomal Q-variants among individuals in the samples.

The distributions of Q-variants, first of all, indicate the frequency of individuals with different numbers of chromosomal Q-HRs in a given sample. At the same time, this indicator reflects the limits of the scope of variability in the number of Q-variants—by this we mean the maximum and minimum values of the number of Q-HRs in certain individuals in the respective populations. This quantitative characteristic of the variability of chromosomal Q-HRs was first presented by us in the comparative description of human populations [Ibraimov et al. 1986, 1990, 1991].

Since there is no information in the literature on the distribution of Q-HRs variants in human populations, I will limit myself to presenting data from our laboratory for some racial/ethnic groups living in different climatic and geographical conditions in Eurasia and Africa, as well as in a sample of mountaineers (Table 2.14.1).

Table 2.14.1. The distribution and mean number of chromosomal Q-HRs per individual in certain samples of Eurasian and African individuals.

Number of Q-HRs	I	II	III	IV	V	VI	VII
0	57	58	94	24	10	1	
1	116	123	221	80	13	1	1
2	141	175	363	141	51	11	21
3	56	99	242	111	85	8	54
4	13	51	137	65	58	13	78
5	2	14	50	23	49	9	79
6			13	5	24	2	55
7			2		3	2	29
8					4	1	6
9							3
							1
Mean number ofQ-HRs	1.63	2.01	2.26	2.44	3.52	3.71	4.68

I – mountaineers (n = 385); II – northern Mongoloids (n = 520); III – highland Mongoloids (n = 1122); IV – Russians in Kyrgyzstan (n = 449); V – steppe Mongoloids (n = 297); VI – Indians of northern India (n = 48); VII – lowland Negroids (n = 327).

This table gives primary data on the distribution of Q-variants in the samples studied by us and also the mean number values of Q-HRs per individual in the relevant populations. This is done in order to more clearly show how the value of this quantitative characteristic of chromosomal Q-HR variants in various populations is formed.

It can also be seen in the table that the mean number value of the Q-variants is close to the modal class of distribution of Q-HRs in human populations. It is noteworthy that populations characterized by low value mean numbers of Q-variants, as a rule, have a narrow range of variability in the distribution of Q-HRs and *vice versa*.

2.15. The frequency of inverted chromosome 3 in human populations

Allderdice [1973] first described the pericentric inversion of the Q-heterochromatin segment of chromosome 3 (*inv 3*) (p15q12) in two phenotypically normal individuals. Soudek et al. [1974] then discovered such an inverted chromosome in three patients with mental retardation. However, the authors, in the same report, suggested (after additional study of the siblings of their patients with pericentric inversion) considering such a variable chromosome 3 (*inv 3*) polymorphic cytogenetic trait. Subsequently, Soudek and Sroka [1978] studied the frequency of *inv 3* chromosomes in 370 mentally retarded patients and 222 intellectually normal individuals. The authors found that, in the first group of individuals, the frequency of *inv 3* was 4.05 %, while in the second it was 4.3 %. The same researchers studied *inv 3* segregation in six families and did not find any significant deviation from the expected Mendelian segregation. Mikelsaar et al. [1978] came to the same conclusions after a comparative study of 102 normal newborns and 45 individuals with mental retardation of Estonian nationality.

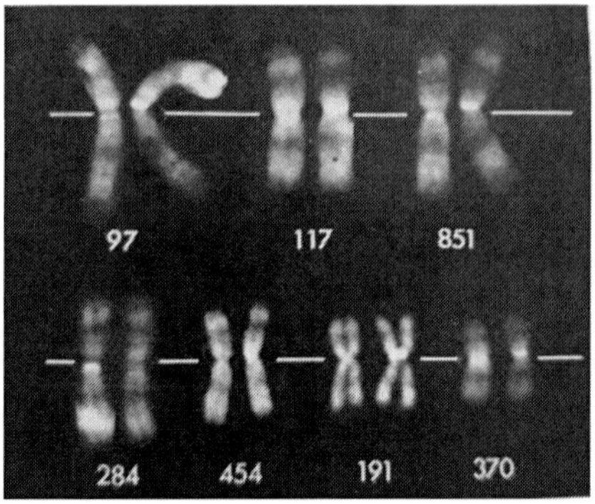

Fig. 2.15.1. Q-stained chromosome 3 pairs from 7 newborn infants. Reallocation of the "bright" band onto p11 is shown on the righthand side of each homologue (reproduced from Lin et al. 1976 with permission of the publisher).

Fogle and McKenzie [1980] studied a single large Negroid family, consisting of 81 members and spanning four generations. They found *inv 3* in 23 individuals; 3 of them had this inversion in homozygous form. The authors found that none of the 23 members of this family, having an *inv 3* chromosome in both hetero and homozygous forms, had any noticeable mental or other pathological abnormalities. Kivi and Mikelsaar [1981], examining 74 women with malignant tumors of the mammary glands and ovaries and 80 healthy individuals of the same sex, found no difference between them (8.1 % and 7.5 %, respectively), which agrees well with data previously obtained in their laboratory when studying the frequency of *inv 3* chromosomes in the Estonian population [Mikelsaar et al. 1978]. To date, the relationship between the pericentric inversion of Q-HRs on chromosome 3 (*inv 3*) with any phenotypic features of carrier individuals has not been established for either the homo or heterozygous form [Soudek and Sroka 1978].

Verma and Dosik [1980b] undertook a special comparative analysis of the conformity of the Q and C-heterochromatic segments of chromosome 3 after Q and C-staining. In some cases, the authors found a complete correspondence between the localization and size of Q and C-variants, that is, intensely or brightly fluorescent 3p11q11 segments gave a dark color after C-staining. However, they observed cases where there was no intense fluorescence on this polymorphic region of chromosome 3, despite the fact that the C-staining revealed the presence of C-heterochromatin and *vice versa*. Based on this observation, the authors concluded that the Q and C-staining of chromosomes did not provide identical information regarding the polymorphism of the heterochromatic region of chromosome 3.

Soudek and Sroka [1978] rightly pointed out that not all authors pay attention to this interesting cytogenetic marker in their population studies. Therefore, in all the populations of Eurasia and Africa studied in our laboratory, the frequency of *inv 3* chromosomes has been carefully recorded. Table 2.15.1 presents data from the literature and the results of our own observations. It is difficult for us to judge the results of other authors, because, as shown in a survey of several authors of population cytogenetic studies made by Soudek and Sroka [1978], many of them did not pay any particular attention to *inv 3*. Nevertheless, from this table it is clear that in all Caucasoid populations, during the examination of which the authors paid attention to *inv 3*, they always found such a chromosome with a frequency ranging from 0.3 to 11 %. However, a comparative analysis of the data obtained by different authors is not possible for the same methodological reasons as described in detail in Section 2.9.

Therefore, in the future, I will confine myself to the results of our laboratory obtained for all three main races: the Mongoloids and Europoids of Eurasia and the Negroids of Africa (Table 2.15.1). Of all the populations we examined, the frequency of this inversion was highest among Russians, amounting to 6.0 %, and, in this sample, it was found even in the homozygous state (0.5 %). Of the eight studied Mongoloid populations of Asia, we met such an inversion only in five, with a frequency ranging from 0.3 to 3.0 %. We have previously explained this fact by the presence of an "impurity" in the Caucasoid gene pool, either due to the initial ethnic composition (Kazakhs and Kyrgyz), or to a certain stage of political history in the past (Mongols, Chukchi, and Yakuts), i.e. the "founder effect" [Ibraimov and Mirrakhimov 1982 a, b, c, 1985; Ibraimov et al. 1982, 1986]. Otherwise, it is difficult to explain why studies of 400 Japanese [Yamada and Hasegawa 1978], 124 Chinese, 120 Khakas, and two populations of Pamir Kyrgyz have not found a single case of such an inversion. In addition, we did not find such an inversion among Negroids of subequatorial Africa. However, the inversion of the Q-heterochromatic segment of chromosome 3 was found in a study of 52 Ethiopians (2.9 %) and the existence of a Caucasoid component in the gene pool is generally recognized.

Table 2.15.1. The frequency of inverted chromosome 3 in human populations.

Place of residence	Frequency in %	Sample size	Race and nationality	Authors
USA, Bronx, NY	0.3	316	Mixed	Muller et al.,1975
Canada, Hamilton	0.75	930	Euopoids (87%)	Lin et al., 1976
Scotland, Edinburgh	1.7	482	Europoids, Scotch	Buckton et al., 1976
Canada, Kingston	4.05	222	Europoids, Anglo-Saxon	Soudek, Sroka, 1978
USSR, Tartu	5-6	102	Europoids, Estonian	Mikelsaar, 1979
USSR, Riga	11.0	140	Europoids, Latvian	Kruminia, Kroshkina, 1982
Northern Finland	3.8	91	Europoids, Finnish	Herva, 1981
Russia, Chukotsk Russia, Yakutsk	0.8 2.5	132 157	Local Arctic type of the Mongoloid race, Chukchi	Personal data Personal data
Southern Kazakhstan	3.0	101	Central Asian type of the Mongoloid race, Khakass	Personal data
Kyrgyzstan	0.3	1319	South Siberian type of the Mongoloid race, Kazakh	Personal data
Turkmenistan	1.7	116	South Siberian type of the Mongoloid race, Kyrgyz	Personal data
Kyrgyzstan	6.0	274	Eurepoids with a Mongoloid admixture, Turkmen Europoids, Russia	Personal data
Mongolia	2.8	72	Central Asian typeof the Mongoloid race, Mongolians	Personal data

Northern China	0.0	124	Eastern Asian type of the Mongoloid race, Chinese	Personal data
Ethiopia	2.9	56	Local Ethiopian race, Ephiopians	Personal data
Angola	0.0	41	Negroids	Personal data
Mozambique	0.0	55	Negoids	Personal data

In our laboratory, Kurmanova [1991] studied 277 mountaineers of Russian nationality, in whom *inv 3* was seen in 23 individuals; its frequency in the sample was comparable to that in the control (6.8 % and 6.0 %, respectively). Thus, according to this quantitative characteristic of the variability of chromosomal Q-HRs, these two Russian samples were not significantly different. In the other two groups of mountaineers (mestizo and mixed groups), *inv 3* was met with a frequency of 4.8 % (3 individuals) and 4.4 % (2 individuals), respectively. Unfortunately, in no case was she was able to conduct a family analysis. However, the presence of *inv 3* in the"mestizo" group can be explained by the fact of having at least one Russian parent. Both individuals with *inv 3* in the mixed group turned out to be Caucasians, which once again testifies to the assumption that *inv 3* is a kind of "Caucasoid" marker.

In connection with the above, we believe that inverted chromosome 3 (meaning the pericentric inversion of its Q-heterochromatin segment) is a kind of "Caucasoid" cytogenetic marker. By the frequency of its occurrence in the sample, one can judge the presence of a European "impurity" in a given population [Ibraimov and Mirrakhimov 1982a, b, c, 1985; Ibraimov et al. 1990, 1991]. Our findings are fully supported by Rossi [1985] in a study of a number of Mexican populations. In this regard, it seems interesting to us to use this cytogenetic marker in ethnic anthropology when studying the intensity of the process of methization in modern society where there is a mixture of different racial, national, and ethnic groups, because we can accurately determine the frequency of *inv 3* in the original populations. Compared to other quantitative characteristics of the Q-variants of chromosome *inv 3*, it has the undoubted advantage that it is almost impossible to make a mistake in its registration, therefore the data of different authors is likely comparable [Ibraimov 2015a].

2.16. The variability of size and fluorescence intensity of chromosomal Q-HRs in human populations

The study of the variability of the size and fluorescence intensity of human chromosomal Q-HRs is an under-developed problem. The need to calculate and register this quantitative characteristic of Q-variants in human karyotypes has been stressed repeatedly in the documentation of the Paris Conference [Paris Conference 1971, Suppl. 1975; ISCN 1978]. However, this document does not provide any specific recommendations for assessing the size of Q-polymorphic segments (Table 2.16.1).

Table 2.16.1. Morphological criteria for semi-quantitative assessment of the variability in size of chromosomal Q-HRs [Paris Conference 1971, 1975].[a)]

Sise level in degrees	Visual criteria for semi-quantitative evaluation of Q-HRs of options under the microscope	
1	Very small	(≤ 0.25 x 18p)
2	Small	(> 0.25 – 0.5 x 18p)
3	Intermediate	(> 0.5 – 0.75 x 18p)
4	Large	(> 0.75 – 1.0 x 18p)
5	Very large	(> 1.0 x 18p)

[a)] As a standard reference within the same cell, the size of the short arm of chromosome 18 (18p) was used.

Later, a number of authors tried to indirectly estimate variations in the size of Q-variants in populations by comparing their size with that of 21q [Müller and Klinger 1976] or 18p [Yamada and Hasegawa 1978] chromosome segments. However, none of the proposed recommendations for estimating the variability in size of Q-heterochromatin segments has become widespread.

Unfortunately, not all authors give data on the quantitative characteristics of Q-variants in their work. If such work is performed, the results obtained should be comparable, to an even lesser extent than the results concerning the Q-HR frequencies in the 12 variable loci of the seven pairs of autosomes and the mean number of Q-variants calculated per individual in the sample.

Initially it was assumed that data on the intensity of fluorescence and the size of Q-variants would make the quantitative characteristic of Q-HRs more objective. To this end, standards have been developed to bring this data to a single "point of reference" and allow, at least, the comparison of results obtained by the same group of researchers at different times. The Paris Conference [1971] recommended the registration of 5 degrees of fluorescence of Q-variants, for each of which an intracellular standard was proposed (Table 2.16.1). However, it quickly became clear that in practice exact adherence to these recommendations was impossible. As a rule, researchers take into account only those Q-variants at the 4th and 5th degrees of fluorescence [1971]. The remaining Q-bands, at degrees 1, 2, and 3 of fluorescence intensity, are accepted as "normal" [Buckton et al. 1976]. In some papers, the 4th and 5th degrees of fluorescence intensity are not highlighted and those bands that are intensely and brightly fluorescent are all described as "intensely fluorescent" [Geraedts and Pearson 1974; Al-Nassar et al. 1981].

In our studies, we followed the recommendations of some Japanese authors, according to which the sizes of Q-HRs on the autosomes were visually divided into five classes under a microscope, comparing them with a short arm of chromosome 18 as follows: Q variants ≤ 0.25 x 18p, were considered very small; > 0.25 - 0.5 x 18p small; - > 0.5 - 0.75 x 18p intermediate; > 0.75 - 1.0 x 18p large; and ≥1.0 x 18p very large (Fig. 2.16.1).

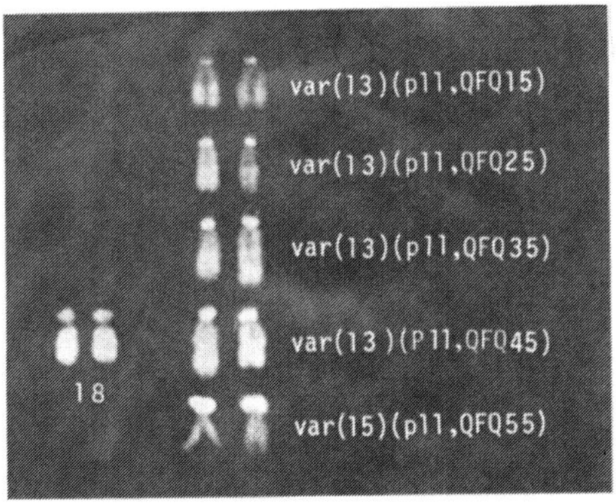

Fig. 2.16.1. Examples of variants showing five levels of difference by comparison to the size of 18p of chromosome 18. The fluorescence intensity was at level 5 in all the Q-HR variants (reproduced from Yamada and Hasegawa 1978, with permission of the publisher).

In all the populations we examined, Q-segment sizes of 4th and 5th degrees of fluorescence intensity fit, mostly (more than 95 %), into classes QFQ 14, QFQ 15, and QFQ. The abbreviation "QFQ" means "Q-bands by fluorescence using quinacrine", as recommended by the Paris Conference [1971, Suppl. 1975]. For example, "QFQ 25" means that after staining the slide with quinacrine in the locus in question, a Q-variant of small size (2) with very bright fluorescence (5) was found. Similar data was obtained by Yamada and Hasegawa [1978] in a study of 400 Japanese in Tokyo and by Herva [1981] in a study of 91 Finns in Northern Finland. The authors noted that, if the short arm (18p) of chromosome 18 is used as a morphological criterion for a visual semi-quantitative estimate of the size of Q-variants, the vast majority of Q-segments correspond to the first two points. In addition, they indicated that the 18p segment in the human karyotype is very large and using it comparatively would make it possible to divide the Q-segments into more fractional semi-quantitative gradations. We have come to the same conclusion after studying more than two dozen human populations [Ibraimov et al. 1981; Ibraimov and Mirrakhimov 1982 a, b, c, 1985; Ibraimov et al. 1982, 1986, 1990, 1991, 1997].

As such, it should be recognized that the problem of accurately determining the sizes of chromosomal Q-polymorphic variants in human populations has not yet been solved and requires further research. The fact is that the interpopulation differences found so far, for example, in terms of the mean number of Q-variants per individual in a population, show only a fraction of the actual heterogeneity of the studied samples. If we had a more accurate method for quantifying the variability in size of Q-HRs, then, perhaps, our attempts to ascertain the possible biological role of chromosomal HRs would advance much further and faster [Ibraimov 2015b].

2.17. The frequency of homo (+/+ and -/-) and heterozygote (+/-) chromosomal Q-HRs in human populations

A number of authors of cytogenetic population studies, provide data on the frequencies of homo and heterozygotes in their samples, along with other quantitative characteristics of Q-HR variability in human populations. As noted above, the terms "homo" (+/+ and -/-) and "heterozygote" (+/-) refer to the presence or absence of Q-variants in homologous chromosomes. These terms do not take into account the fact that Q-HRs themselves vary continuously in both size and fluorescence intensity, that is, in the characteristics by which they differ morphologically from other fluorescent chromosome segments detected by Q-staining [Paris Conference 1971; Ibraimov and Mirrakhimov 1985].

It should be noted that not all authors of population studies investigate the frequency of homo and heterozygous Q-variants in terms of the Hardy-Weinberg law [McKenzie and Lubs 1975; Schwinger and Wehner 1976; Lin et al. 1976; Yamada and Hasegawa 1978; Herva 1981]. Nevertheless, quite a few works have been published in which the authors analyze, in detail, the correspondences between the observed and expected frequencies of homo and heterozygotes according to the Hardy-Weinberg law [Schnedl 1971; Geraedts and Pearson 1974; Mikelsaar et al. 1974, 1975; Muller et al. 1975; Buckton et al. 1976; van Dyke et al. 1976; Mikelsaar 1981; Tupitsina and Stobetsky 1980; Nazarenko 1987; Al-Nassar et al. 1981; Ibraimov et al. 1982, 1986; Ibraimov and Mirrakhimov 1982 a, b, c, 1985]. In most cases, such a match has indeed been found. However, a number of authors have found some discrepancies between the observed and expected frequencies in some of the Q-polymorphic loci of the seven autosomes, which may have been caused by changes in the segregation relations of chromosomal Q-HR variants in progeny [Michelsaar 1981; Tupitsyna and Stobetsky 1980; Nazarenko 1987; Geraedts and Pearson 1974; Mikelsaar et al. 1974, 1975; Müller et al. 1975; Robinson et al. 1976].

Some authors have attempted to analyze the correspondence of the observed frequencies of homo and heterozygotes in a sample to the theoretically expected correspondence according to the Hardy-Weinberg ratio to confirm hypotheses about the possible selective value of Q-chromosomal Q-HRs of particular localizations. At the same time, cases of statistically significant deviations of observed frequencies from expected ones, if any, relate, as a rule, to one or two loci; no author has described a simultaneous divergence in all loci. Failure to comply with the Hardy-Weinberg ratio has been noted for different loci in the work of different researchers and even in studies produced by a single group of authors. Mikelsaar et al. [1975] noted that, with the exception of 4 cen, 13 cen, 13s, and 15s (13p13 and 15p13), if we use the adopted designation system, all other loci detected deviations from the expected frequencies.

Analyzing the correspondence of the observed frequencies of homo and heterozygotes to those expected based on the observance of the Hardy-Weinberg ratio, in the overwhelming majority of cases the authors have come to the conclusion that there are no significant deviations. It should be noted that, with respect to Q-variants, the concepts of "homozygote" and "heterozygote" are largely formal, concerning individuals who have (+/+) or do not have (-/-) Q-HRs on both homologous chromosomes, and heterozygote individuals who have Q-HRs on only one of the homologues (+/-). This does not take into account the degree of fluorescence of Q-variants or their size, i.e., despite the existence of a significant quantitative component in Q-HR variability, even minimal quantitative restrictions are not introduced apart from one: the fluorescence is stronger than that of two sharp Q+ positive bands of chromosome 9 (9q), which naturally affects the content of the terms “homozygote” and “heterozygote”.

From the point of view of formal genetics, all Q-variants resulting in fluorescence of degrees 4 and 5 of this locus after staining can be considered to constitute a single allelic variant. The same applies to the “non-fluorescent” variants. All of them can be considered to be alternative fluorescent alleles. Using this method of calculation and registration, the chromosomal Q-HRs system appears as a system of two sets of isoalleles. Consequently, we have to check the subordination of the variability of this system to classical laws, in particular to the Hardy-Weinberg rule. In this case, we can consider it a diallele system in which one of the alleles is constituted on “the presence” of Q-variants in any of the 12 potentially variable loci on the seven autosomes and the other is the “absence” of Q-variants in this locus.

When studying the segregation behavior of Q-variants, a significant prevalence of heterozygotes for autosomes 13 and 15 was found in Dutch populations [Geraedts and Pearson 1974]. The authors believe that the most likely cause of this phenomenon is the selective advantage of heterozygotes over both homozygote forms, although they do not exclude the influence of the sample size and methodological errors. In addition, on the basis of an analysis of the segregation behavior of Q-variants, they found that an excess of heterozygotes occurs “from the maternal side—in the sons and from the father—in the daughters”. A similar fact was established by Robinson et al. [1976] for brightly fluorescent Q-variants. Müller et al. [1975], examining the karyotypes of 376 newborns, found that the Hardy-Weinberg ratio is not subject to Q-HRs localized on chromosomes 3 and 22, for which a “homozygous deficit” was found, while there are statistically significant other loci correspondence between observed frequencies and expected ones. The inconsistencies found by this group of authors, however, were not large.

Buckton et al. [1976] did not find any deviations from the Hardy-Weinberg ratio on any locus in a survey of fairly large samples from three age groups of the Scottish population and concluded that there was a genetic equilibrium in the studied populations with respect to fluorescent polymorphism.

Mikelsaar et al. [1974], in examining a normal adult population of Estonians, found differences between the sexes in the brightly fluorescent Q-HRs of chromosome 3, seeing more homozygotes with brightly fluorescence Q-HRs in men and heterozygotes in women. They also [Mikelsaar et al. 1975] found a deviation of the observed frequencies at loci 14p13, 21p13, 22p11, and 22p13 in the prevalence of homozygotes and lack of heterozygotes in children with Down's syndrome and mental retardation of unknown etiology. The authors suggested three possible explanations for this fact: 1) the dependence of the fluorescence of chromosome segments on factors associated with sex (hormones); 2) sex-dependent selection of homo and heterozygotes; and 3) the existence of non-random segregation in meiosis of the brightly fluorescent band of chromosome 3. However, none of these explanations has been confirmed.

Many authors have plausibly suggested that an excess of heterozygotes in a particular Q-variable locus can be explained by selection in their favor over homozygotes, or by anomalous segregation of chromosomes carrying Q-HRs in meiosis. There are also reports of the possible effects of parental age on the Q-HR segregation processes [Mikelsaar 1981; Tupitsina and Stobetsky 1980; Mikelsaar et al. 1974, 1975; Geraedts and Pearson 1974; Robinson et al. 1976].

In turn, we [Ibraimov and Mirrakhimov 1982 a, b, c; Ibraimov 1983; Ibraimov et al. 1982, 1986] did not find any statistically significant differences in a survey of more than two dozen populations belonging to three racial groups, and even in those cases when sample sizes were small (not exceeding 50 individuals) or where populations were subject to selection pressure in the extreme conditions of high altitude in the Pamir Mountains and Tien Shan, as well as in the Far North.

Nazarenko [1987] found statistically significant deviations of observed Q-HR frequencies from those expected in a group of northern Khanty men with a deleted Q-HR block on the long arm of the Y chromosome, at locus *3 cen*, and this deviation was due to excess heterozygotes; in the other 11 loci examined there was no such deviation. These data served to substantiate the hypothesis of the existence of "reciprocal variability in the content of certain heterochromatin fractions, as a result of which their level in the human genome is maintained within certain optimal limits". Considering the loss to be a result of the deletion of a certain fraction of heterochromatin in the Y chromosome "predisposes" it to the non-random segregation of carrier Q-HRs of chromosome 3 in meiosis—the author sees this as a violation of the Hardy-Weinberg rule.

Studying a group of 50 married couples with a history of spontaneous abortion of unknown etiology, Kruminia et al. [1987] found differences in the frequency of occurrence in the pathology groups of the control of "homo and heterozygotes on the basis of bright fluorescence" for most of the loci under consideration (except for chromosome 22); moreover, compared to the control, in the pathology group the frequency of homozygotes (-/-) increased and the frequency of homozygotes (+/+) reduced. The authors believe that this observation favors their assumption of the negative effect on the body of an elevated Q-HR content in the karyotype. As such, when considering the work of various groups of authors, it is not possible to trace an unambiguous trend.

Ginsburg et al. [1991] studied the statistical side of the issue. Leaving aside the question of the uniqueness of the interpretation of a detected violation of the Hardy-Weinberg rule, the authors turned to the statistical procedure by which this violation can be detected. At the same time, they showed that if the long-established $\chi 2$ test for samples of the most frequent volume (25-200 observations) is used, and the Hardy-Weinberg distribution hypothesis is correct, the researcher will nevertheless reject it more often than the lack of the gene in human populations, and, in some cases, the error of the first kind reaches a value that makes the study itself incorrect.

Based on the above, it may be assumed that the current use of the frequency of homo and heterozygotes of chromosomal Q-HR variants in human populations is not justified. This method, being rather formal in nature, has not provided anything new in our understanding of the biological nature of the wide quantitative variability of chromosomal Q-HRs in human populations. From the point of view of cytogenetics, this idea has most clearly been expressed by Jacobs [1977], who noted that it is too early to assess the compliance of observed and expected homo and heterozygotes according to the Hardy-Weinberg law, since the existing methods for calculation and registration of chromosomal Q-variants are not perfect due to the continuous nature of the size distribution of heterochromatic regions in human populations.

Our attitude to the use of this quantitative characteristic of chromosomal Q-HRs in human populations can be reduced primarily to the following: until we get strong evidence that Q-variants localized in the 12 potentially polymorphic loci of the seven pairs of human autosomes are fundamentally different structural and genetic elements, the use of the Hardy-Weinberg ratio remains, basically, an exercise in solving certain problems of applied mathematics with no significant biological content. Mayr [2002] has specifically discussed this issue, saying: "They expressed this as a mathematical formula, which is a reapplication of a mathematical law, the binomial expansion. Being a strictly mathematical solution, it is not a *biological law*".

In general, it may be considered that the available observations of some deviations in the segregation of Q-variants in families or on the discrepancy between the observed and expected frequencies of homo and heterozygotes in human populations do not yet require a particular biological interpretation [Ibraimov 2015a.b].

2.18. The variability in size of the Q-heterochromatin segment of the Y chromosome in human populations

Significant interindividual variability in the length of the Y chromosome in human populations was noted long before the development of suitable methods for the differential staining of chromosomes. The Denver Conference on the standardization of the nomenclature of human mitotic chromosomes [Denver Conference 1960] highlighted the variability in the size of the long arm of the Y chromosome in different individuals and different ethnic groups. It soon became apparent that variable Y chromosomes are highly stable morphological features of the human karyotype and are inherited [Bishop et al. 1962; de la Chapelle et al. 1963; McKenzie et al. 1972].

In the same period, a variety of structural Y chromosome aberrations, such as isochromosomes [Jacobs and Ross 1966; Ferguson-Smith et al. 1969], dicentrics [McIlree et al. 1966; Armendares et al. 1983; Sieberg et al. 1973], rings [Chandley and Edmond 1971; Ruthner and Golob 1974], translocations [Noel et al. 1971; Krmpotic et al. 1972], inversions [Solomon et al. 1964; Jacobs and Ross 1966; McIlree et al. 1966], and deletions [Conen et al. 1961; Bengtsson et al. 1974; Langmaid and Laurence 1974] in both phenotypically abnormal and normal men were identified. Here, I will focus mainly on the variability in the length of the Q-heterochromatin segment of the Y chromosome in different human populations.

Usually, there are short and long Y chromosomes, depending on the value of the Y/F index. Y chromosomes that are equal or smaller in length than the chromosomes of group G (pairs 21 and 22) are considered short or small. Most often, population studies record the frequency of long Y chromosomes. The long, or large, Y chromosome, as defined by Court Brown [1967], must be equal to or greater than the chromosomes of the F group (pairs 19 and 20).

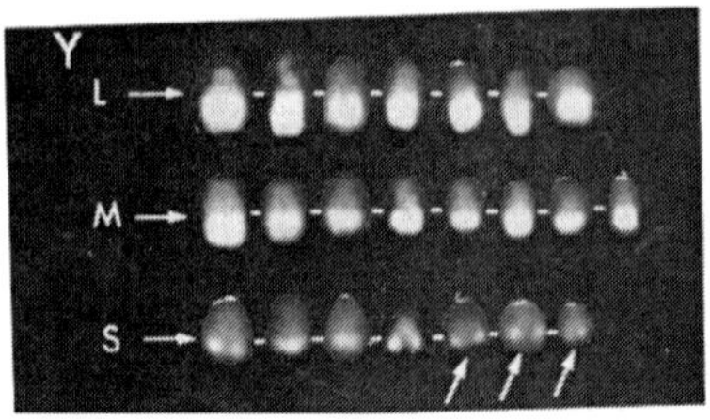

Fig. 2.18.1. Variable bright fluorescent regions on the distal long arm of Y chromosomes from different male newborn infants. Y chromosomes with very small bright fluorescent region are indicated by arrows (reproduced from Lin et al. 1976 with permission of the publisher).

Since the very beginning of their detection, variable Y chromosomes have attracted the attention of cytogeneticists. Despite the fact that the nature of this variability is still not fully understood, the main directions of the search should still briefly be considered. Bishop et al. [1962] concluded that a change in the length of the Y chromosome is, partially or completely, related to the degree of contraction of the mitotic chromosomes during cell division. Gripenberg [1965] argued that an increase in the length of the Y chromosome reflects the presence of additional chromosomal material, since he was able to observe that long Y chromosomes have two secondary costrictions on a long arm. However, Wehlström and de la Chapelle [1963], and later Kikuchi and Sandberg [1965], compared the number of H3-thymidine grains (a radioactive label) included in the cell of men with very long and normal Y chromosomes and did not find any significant differences between them. Wehlström [1971] reported that duplication of the distal part of the long arm of the Y chromosome is one of the possible mechanisms driving the occurrence of long Y chromosomes.

Zech [1969] discovered that, after staining human chromosome preparations with quinacrine mustard, the distal portion of the long arm (q12 segment) of the Y chromosome gives the brightest fluorescence. This valuable observation marked the beginning of a qualitatively new stage in the study of Y chromosome polymorphism in human populations. It soon became clear that the human Y chromosome consists of two parts: 1) the non-fluorescent—the proximal part of the long arm, the centromeric region and the short arm of the Y chromosome; and 2) the fluorescent—the distal half of the long arm. In weakly spiralized chromosomes, the Y chromosome can have two or more brightly fluorescent segments separated by weakly fluorescent segments [Soudek et al. 1973; Yamada and Hasegawa 1978; Zakharov et al. 1982]. The increase in the size of the brightly fluorescent region on the chromosome (1/2 of the arm length and more) is most often associated with an increase in the number of Q-segments. When reducing the size of the brightly fluorescent area (1/4 of the shoulder length and less), its division into segments is not observed.

The first studies showed that the variability of the long arm of the Y chromosome is mainly associated with changes in the size of the brightly fluorescent segment, while the length of the non-fluorescent segment remains unchanged [Bobrow et al. 1971b; Ricci et al. 1971; Laberge and Gagne 1971]. However, later on, using more accurate quantitative methods of analysis, it was shown that the variability of the length of the Y chromosome is influenced by the dimensions of not only the brightly fluorescent, but also the non-fluorescent regions of the long arm [Schnedl 1971; Tishler et al. 1972; Soudek and Sroka 1978; Yamada and Hasegawa 1978; Verma et al. 1978b, 1983; Skawinski and Parcheta 1984].

It is widely believed that short Y chromosomes result from deletion [Conen et al. 1961; Muldal and Ockey 1961; Genest et al. 1970]. Muldal and Ockey [1961] came to this conclusion due to the lack of the distal part of the long arm in the short Y chromosome. Wehlström [1971] observed a very small Y chromosome, the distal part of which did not fluoresce after Q-staining. A phenotypically normal male was described in which the q12 segment of the Y chromosome did not produce bright fluorescence after Q-staining; C-staining, however, revealed the presence of C-heterochromatin on this segment [Soudek and Laraya 1976]. On the other hand, Fonatsh et al. [1977] described a man with a 45, X/46, XY karyotype, who also did not have Q-heterochromatin material on the Yq12 segment, but did not detect the presence of C-heterochromatin using C-staining. Based on this observation, the authors concluded that negative C-bands on a chromosome correspond to areas with no bright fluorescence. Finally, quite convincing information has been accumulated indicating that the Q-heterochromatic segment of the Y chromosome contains a large number of highly repetitive DNA sequences that do not transcribe messenger RNA [Pardue and Gall 1970; Arrighi and Hsu 1971; Yunis et al. 1971].

Thus, the set of information accumulated to date about the variability of the Y chromosome in humans suggests that it should be considered a genetically polymorphic trait, because: 1) it is found in populations at a relatively high frequency; 2) it is usually associated with a normal phenotype; and 3) it is always inherited generationally [Sandberg (Ed). The Y chromosome: Part A and B, 1985].

Cohen et al. [1966] first reported on the existence of interracial differences in the frequency of occurrence of long Y chromosomes. In particular, the authors found that Japanese men with long Y chromosomes were more common than Caucasians, Negroids, and American Indians. Angell [1973] showed that the length of the Y chromosome in Australian aborigines is significantly less than in Caucasians. Ghosh and Singh [1975] presented data on the heterogeneity of Indian populations and found that, in particular, among Rajputs the Y chromosome, is significantly longer, on average, than Punjabis.

Lubs and Patil [1975] suggested that there was a definite gradient in the variability of the length of the Y chromosome in the Europoids of Europe, going from north to south. In particular, in men of Mediterranean origin, long Y chromosomes are most common. Matsuura et al. [1979] found that the Y chromosomes of Mongoloids (mainly Japanese) are significantly longer than those of the Caucasians, thereby confirming the preliminary observations of Cohen et al. [1966]. In addition, the authors found that, in a sample of Filipinos, the size of the Y chromosome was the same as that of Mongoloids, whereas individuals of Polynesian origin had Y chromosomes much smaller than Europoids. Based on these observations, Pacific aborigines were divided into three groups: 1) Australian aborigines with the smallest Y chromosomes; 2) a Polynesian group characterized by a medium-sized Y chromosome; and 3) Filipinos and eastern Mongoloids with large Y chromosomes.

Undoubtedly, the size of the Q-HR of the Y chromosome segment is one of the most interesting characteristics of population variability concerning the number of chromosomal Q-HRs and its possibilities are far from having been completely understood. In the literature, there are numerous data on the frequency of long Y chromosomes in human populations, as identified by both routine and differential staining methods (Table 2.18.1). These methods usually do not give comparable results and even data obtained using the differential staining of chromosomes should be assessed critically if they are not based on adequate quantitative measurements (see below), since only a few studies have used uniform criteria to assess the variability of Y chromosome size.

Table 2.18.1. Frequency of the long Y chromosome in human populations.

Population	Number of man	Long Y chromosomes (in %)	Authors
Europoids (USA)	20	5,0	Cohen et al., 1966
Asian Indians	20	10,0	Cohen et al., 1966
(USA)	20	10,0	Cohen et al., 1966
Blacks (USA)	20	15,0	Cohen et al., 1966
Jews (USA)	20	45,0	Cohen et al., 1966
Japanese (USA)	259	5,8 a)	Cohen et al., 1966
Newborns	207	1,5 a)	Court Brown, 1967
(Jerusalim)	30	13,4	Unnerus et al.,
Europoids	2444	6,6	1967
(Scotland)	43	18,6 a)	Lubs, Ruddle,
Europoids (Finland)	411	14,8	1971
Newborns (USA)	74	9,4 a)	LaTorre et al.,
Europoids (Spain)	71	2,9	1970
Newborns (USA)	147	2,1	Lubs, Ruddle,
Europoids	3468	1,0 a)	1971
(Sweden)	140	1,4	Lins, Sunderquist,
Poles (Poland)	2615	1,0 a)	1971
Newborns	238	12,2	Huebner, 1971
(Germany)	52	3,8	Zankl, Zang, 1971
Newborns (Canada)	1303	1,5	
Newborns	7176	1,0	Hamerton et al.,
(Denmark)	5661	1,0	197
Newborns	100	5,0	Nielsen, Friedrich,
(Denmark)	100	3,0	1972
Newborns (Canada)	175	15,9	Friedrich, Nielsen,
Estonia (USSR)	61	26,2	1972
Newborns (USSR)	28	28,6	Soudek et al., 1973
Newborns (Canada)	20	25,0	Mikelsaar et al., 19
Newborns	12	8,3	Bochkov et al.,
(Denmark)	493	9,5	1974
Rudjputs (India)	60	5,0	Hamerton et al., 19
Pendjubs (India)	409	60,9	Nielsen, Sillesen,
Newborns (USA)	21	69,1	1975
Koreans (USA)	19	26,3	Ghosh, Singh,
Chinese (USA)	60	18,3	1975
Japanese (USA)	50	10,0	Ghosh, Singh,
Europoids (USA)	60	10,0	1975
Newborns (Canada)	70	24,3	Muller et al., 1975
Europoids (USA)	126	38,9 b)	Park, 1976
Japanese (Tokyo)	106	28,3 b)	Park, 1976

Arabs, Aimen	55	36,4 b)	Park, 1976
(Kuweit)	148	15,6 b)	Park, 1976
Arabs, Suluba	54	37,1 b)	Lin et al, 1976
(Kuweit)	52	51,9 b)	Verma et al., 1978
Arabs of city of	49	22,4 b)	Yamada et al.,
Kuweit	45	8,9 b)	1981
Newborns (Finland)	132	4,5 b)	Al-Nassar et al.,
Blacks (USA)	40	15,0 b)	1978
Eastern Indians	51	26,4 b)	Al-Nassar et al.,
(USA)	34	11,8 b)	1978
Kyrgyz (USSR)	22	9,1 b)	Al-Nassar et al.,
Russians (USSR)	33	21,2 b)	1978
Chukchi (USSR)	46	2,2 b)	Herva, 1981
Mozambique			Verma et al., 1982
Yemen			Verma et al., 1983
Chinese (USSR)			Own data
Khakass (USSR)			- // -
Turkmen (USSR)			- // -
Angola			- // -
Kazakh (USSR)			- // -
Ephiopia			- // -
Zimbabwe			- // -
Khants (USSR)			- // -
Selkups (USSR)			- // -
Yakuts (USSR)			- // -
			- // -
			- // -
			- // -
			- // -
			- // -

a) $Y \geq F$, according to the criterion of Court Brown [1967];
b) Unpublished data. Here, the frequencies of long Y chromosomes are given according to the Court Brown criterion [1967].

Nevertheless, in commenting on the data in Table 2.18.1, a number of circumstances should be noted. First, four independent observations have revealed that the Japanese can be characterized as having the highest frequency of large Y chromosomes [Conen et al. 1961; Park 1976; Matsuura et al. 1979; Yamada et al. 1981]. Secondly, based on extensive research conducted on the very different populations of Eurasia and Africa, it may be assumed that the frequency of the spread of large Y chromosomes, in general, is determined by the ancestor effect and, apparently, does not have a noticeable geographical, ecological, or other gradient. Lastly, of all the human populations studied to date, the largest Y chromosomes have been found in populations of East Asia (Japanese,

Koreans, and Chinese) (for more details, see Ibraimov and Mirrakhimov 1985).

Obtaining more accurate information about the polymorphism of the Y chromosome in populations has largely been delayed due to the lack of accurate methods for quantifying its euchromatin and heterochromatin segments. All currently existing methods of semi-quantitative assessment of the length of the Y chromosome do not give reproducible and comparable results. Of all the known methods for quantifying Y chromosome polymorphism [Lin et al. 1975; Muller et al. 1975; Yamada and Hasegawa 1978; Verma et al. 1978b], in our opinion, the most suitable is the chromosome length measurement system used by the Verma group in Brooklyn, New York, USA.

Verma et al. [1978b, 1983] published a series of reports on the accurate quantitative measurement of Y chromosome length in 60 Caucasians, 60 Negroids, and 70 Amerindians. The preparation of chromosome slides, the Q-staining technique, microscopy, and photomicrography were all carried out using generally accepted methods. Five metaphase plates from each individual were selected for measurement. Quantitative measurements were carried out directly on the negatives, because, as the authors showed, much of the information is lost when working from photographic prints [Verma and Dosik 1976].

In the procedure, the length of all the chromosomes of group F (19 and 20), the total length of the Y chromosome, and the length of the brightly fluorescent (f) and non-fluorescent (nf) segments of the Y chromosome of the same metaphase plate are all measured. Based on these measurements, the ratio (or index) of Y/ F, f/F, and nf/F for each cell is determined and the average value for five cells is then calculated. F is equal to the average length of chromosomes 19 and 20. In order to determine the "functional relationship" of variables, we analyze the regression coefficient. The regression coefficient analysis ***r*** is used to investigate the degree to which two variables will vary together. Based on the Y/F index of the length Y of the human chromosome, they are divided into five conditional discrete classes—very small, small, medium, and very large—corresponding to the following numerical intervals: <0.80; 0.81-0.94; 0.95-1.09; 1.10-1.23; and >1.23, respectively. To study the variability of brightly fluorescent and non-fluorescent Y chromosome segments, the f/F and n/F indices are calculated. If the n/F segment of the Y chromosome does not change in size, then the nf/F index must remain constant, since the length of the chromosomes of the F group is always constant [Paris Conference 1971; Ledley et al. 1972]. The functional relationship between the two variables (f/F versus Y/F) is calculated based on ab analysis of the regression coefficient and the regression line.

Using such a quantitative analysis, Verma et al. [1978b] were able to discover that the length of the Y chromosome depends on the size of both the f and nf segments. It was previously thought that the polymorphism of the Y chromosome depends mainly on the variability of the brightly fluorescent q12 segment of the long arm [Bobrow et al. 1971b; Laberge and Gagne 1971]. No links between the values of the f and nf segments were found, in other words, they vary independently of each other. In addition, the Verma group was able to show that there are significant quantitative differences in the distribution of the Y/F index among American Indians, Caucasians, and Negroids. In particular, the Y chromosomes examined by the authors were longer than in other samples ($P < 0.05$), and, interestingly, the increase in the size of the long arm of the Y chromosome, in all cases, was primarily due to the elongation of the non-fluorescent segment (Yq11). Moreover, the authors emphasized that the content of such common definitions as "medium" or "large" for Y chromosomes varies considerably from one race to another [Verma et al. 1983].

Analyzing in detail the studies of Verma and his colleagues on Y chromosome polymorphism in human populations, I would like to emphasize that only on the basis of such quantitative measurements can we obtain comparable and reproducible results, without which a comparative analysis of the cytogenetic population data produced by various authors is unthinkable. Therefore, the data presented in Table 2.18.1 should be considered critically before methodically correct quantitative measurements of Y chromosome length in these populations are carried out.

The range of studies related to the study of Y chromosome polymorphism is very wide and cannot be covered within the framework of a single monograph. The first international report in this area was published in two volumes devoted to different aspects of the Y chromosome in humans and animals and summarized the results of almost all the studies conducted in various laboratories around the world [Sandberg (Ed). The Y chromosome: Pt. A and B, 1985]. The brief information outlined here does not exhaust all aspects of the variability of the Q-heterochromatin segment of the Y chromosome in humans. Within the framework of the problem discussed here, I will return to this aspect in Section 2.20.8.

2.19. Inheritance and stability of chromosomal Q-HRs

Issues of inheritance relating to human chromosomal Q-HR variants detected by Q-staining touch on a number of aspects. First of all, we may ask the question: are these variable structures of chromosomes inherited according to known Mendelian laws?

Researchers have long been interested in the inheritance of extreme variants of heterochromatic bands of chromosomes. It has been shown that the ratio of chromosomal extreme and "normal" variants in progeny is close to the expected value (0.5) [de la Chapelle et al. 1974; Nielsen et al. 1974].

A number of studies have been devoted to studying the inheritance of chromosomal Q-HRs that do not have extreme variants. Phillips [1977] studied the pattern of inheritance of Q and C-variants in 36 individuals in three unrelated families and found that 50 % of descendants inherited chromosomes with polymorphic variants. Robinson et al. [1976] investigated the segregation pattern of polymorphic chromosomes in 32 families and found, in general, full compliance between the observed data and the expectations of Mendelian laws. In particular, the authors noted that some of the apparent deviations were due to errors in calculation and registration of Q-variants. Other researchers have come to the same conclusion [McKenzie and Lubs 1975; Tupitsyna and Stobetsky 1980].

Geraedts and Pearson [1974], in studying the segregation of Q-variants, found that, at least for the 13th pair, there are "too many" heterozygotes and "too few" homozygotes. The authors commented that "a greater number of heterozygotes are transmitted from the maternal side to sons and from the father's side to daughters". Robinson et al. [1976] noted the predominant inheritance of bright segments by sons from mothers and daughters from fathers.

Fogle and McKenzie [1980] examined a large Negroid family consisting of 81 individuals, looking at Q and C-polymorphic variants over four generations. When summing up their data on chromosomes 3, 4, and 13, the authors found that homologues with Q-variants significantly ($P < 0.05$) more often pass to offspring than homologues without Q-heterochromain. By combining their data with data from Robinson et al. [1976] and Phillips [1977], the authors concluded that non-random segregation ratios statistically significantly exist for chromosomes 3, 4, and 13 [Fogle and McKenzie 1980]. Mikelsaar [1981] investigated 102 married couples, each of whom had one child. Analysis of these marriages (+/+ x +/-) and (+/- x -/-) showed a significant dependence of the segregation ratios of Q-variants on the age of parents in three segments

(4p11q11, 13p11, and 21p13) of 12 polymorphic segments of seven autosomes. Commenting on the results of his research, Mikelsaar [1981] noted that deviations in the segregation ratios of Q-variants obviously do exist. He argues that this indicates either the presence of selection aimed at the signs determined by Q-polymorphic segments, or non-random segregation of Q-variants in meiosis.

However, it is still premature to talk about the actual prevalence of heterozygotes, and, even more so, to explain this circumstance by some selective process. It is possible that the above-noted deviations of the observed data from those expected according to Mendel's law can be explained by the well-known methodological difficulties in identifying Q-variants, as well as by the small size of the study samples. In addition, with respect to the variability of chromosomal Q-HRs of homo (+/+ and -/-) and heterozygosity (+/-), this usually refers to the presence or absence of Q-variants at the 4th and 5th degrees of fluorescence intensity on the homologues. At the same time, if we are talking about homozygosity (+/+) of one or another Q-polymorphic autosome, we are referring to the homology of homologues only by localization and less often by fluorescence intensity, not to mention the size or other characteristics of the Q-variants of chromosomes (see also Section 2.10).

The second aspect of the problem concerns the stability of Q-variants in the process of their inheritance. Jacobs et al. [1975], Robinson et al. [1976], van Dyke et al. [1977], and Carnevale et al. [1976] did not find a single case of newly formed Q-variants in a study that looked at 74 families and 23 pairs of monozygotic twins. Studies on mono and dizygotic twins have shown that Q-polymorphic variants are a stable and inherited feature of chromosomes, since differences are found only between dizygotic twins [van Dyke et al. 1977; McKenzie et al. 1972]. However, in the literature there are reports on the registration of "newly formed" polymorphic variants of chromosomes in descendants of families whose biological relationship with their surveyed parents was confirmed not just by studying blood groups and other well-known markers, but also by analyzing the inheritance of Q and C-variants themselves [Craig-Holmes et al. 1975; Nakagome et al. 1977; Phillips 1977; Madan 1978; Mikelsaar 1981].

The third aspect of the problem concerns the stability of chromosomal Q-variants in human ontogenesis. This aspect has not yet been studied. However, some information was obtained in a comparative study of individuals from different age groups, although no one has yet been able to follow changes in certain variants of Q-HRs over an extended timeframe (at least decades) in the same individuals. Buckton et al. [1976], in studying three age samples from one ethnic group, found a slight decrease with age in the mean number of Q-variants per individual. This value was 4.2 in newborns, 3.9 in 14-year-old boys and girls, and 2.9 in older people (65 years and older) (for more details, see Section 2.20.4).

Mikelsaar [1979], in studying chromosomal Q polymorphism in the Estonian population, approached this problem from a completely new perspective. The author analyzed the dependence of the mean quantity of brightly fluorescent segments per cell, not on the age of individuals, but on the year of birth. The conclusion was reached that individuals born during the war (1941-1945) differed from individuals born before and after it, according to the mean quantity of brightly fluorescent Q-bands per cell—for women it is higher, and for men it is lower than for those born before and after the war. In connection with this circumstance, the author believes that "severe external conditions, in particular, starvation, had a direct effect on the chromosomes of the fetus or child, either caused a selection aimed at any significant signs of the body, or caused directional changes in the segregation ratios of Q-variants, perhaps through hormonal shifts in the parents' bodies". However, neither the author nor other researchers have returned to this question.

2.20. Regularity in the distribution of chromosomal Q-HRs in human populations

I don't know anything, but I do know that everything is interesting if you go into it deeply enough.
—R. Feinman

2.20.1. The frequency distribution of chromosomal Q-HRs in human populations

The frequency of Q-variants in 12 polymorphic loci of seven autosomes is a generally accepted quantitative characteristic of the variability of chromosomal Q-HRs in the genome of the human population (see Section 2.12). Speaking of the existence of interpopulation differences, the researchers, first of all, have in mind this circumstance, even in those cases when they do not find a difference in the mean number of Q-HRs per individual in a population. Table 2.20.1.1 presents the results of a comparative analysis of human populations examined by us in terms of the frequencies of Q-variants in seven Q-polymorphic autosomes.

Table 2.20.1.1. The frequency of Q-HRs in seven Q-polymorphic autosomes in human populations living in Eurasia and Africa.

Location of Q-HRs	I (n = 520)	II (n = 1122)	III (n = 449)	IV (n = 297)	V (n = 48)	VI (n = 327)
3	358 (0.344)*	759 (0.354)	378 (0.420)	236 (0.397)	53 (0.552)	425 (0.649)
	34.3**	31.0	34.4	22.6	29.8	27.8
4	32 (0.031)	130 (0.058)	29 (0.022)	16 (0.027)	5 (0.052)	18 (0.027)
	3.1	5.0	1.8	1.5	2.8	1.2
13	332 (0.319)	769 (0.343)	379 (0.422)	309 (0.520)	55 (0.573)	573 (0.821)
	31.8	30.0	34.4	29.6	30.9	35.1
14	63 (0.060)	113 (0.059)	69 (0.077)	93 (0.156)	10 (0.104)	112 (0.171)
	6.0	5.2	6.3	8.9	5.6	7.3
15	86 (0.083)	262 (0.117)	86 (0.094)	140 (0.235)	24 (0.250)	147 (0.224)
	8.2	10.2	7.7	13.4	13.5	9.6
21	125 (0.120)	260 (0.116)	105 (0.116)	135 (0.230)	18 (0.188)	155 (0.237)
	12.0	10.1	9.5	13.1	10.1	10.1
22	48 (0.046)	214 (0.095)	64 (0.071)	113 (0.190)	13 (0.135)	136 (0.207)
	4.6	8.3	5.8	10.8	7.3	8.9
Total	1044	2563	1100	1044	178	1530
Mean number of Q-HRs	2.1	2.28	2.45	3.52	3.71	4.68

n = sample size; * Q-HR frequency of the chromosomes analyzed; ** Q-HR frequency as a percentage of the overall number of chromosomal Q-HRs.
I – northern Mongoloids of Siberia; II – highland Mongoloids of the Pamirs and Tien-Shan; II – steppe Mongoloids of Central Asia; IV – Russians; V – Indians of northern India; VI – lowland Negroids of subequatorial Africa.

Unfortunately, the use of data from other authors to analyze the nature of interpopulation differences in the autosomal Q-HR frequency distribution pattern is not possible due to the methodological problems outlined in Section 2.9. Therefore, in the future, we will restrict ourselves to analyzing our own results. Even cursory analyses of Table 2.20.1.1 convincingly suggest that Q-HRs are distributed across the potentially Q-polymorphic loci of seven autosomes by more than chance. In fact, more than half of the Q-variants are localized on autosomes 3 and 13, while the rest are more or less evenly distributed on autosomes 4, 14, 15, 21, and 22.

When comparing these two types of Q-HR frequency (* and **), a question arises regarding the nature of the underlying interpopulation differences: a) different degrees of polymorphism of 3 cen, 4 cen, 13p11, 13p13, 14p11, 14p13, 15p11, 15p13, 21p11, 21p13, 22p11, and 22p13 loci; or b) proportional (i.e., depending on the degree their polymorphism) increase or decrease in the absolute number of Q-variants in these loci in a given population [Ibraimov 1983b, 1993]? Such a question, at first glance, seems far fetched. The fact is that, if not the causes, then at least some quantitative patterns of interpopulation differences in the distribution of Q-variants in seven autosomes are extremely important. This means that we must develop a clear idea of what, in fact, makes up the heterogeneity of human populations in terms of chromosomal Q-HRs. If we simplify the situation, the choice will be limited. Namely, either we allow for the selective mutability of these 12 polymorphic loci in different populations, recognizing their locus specificity, or we admit their principal structural and functional homogeneity in the rank of the system, meaning that chromosomal Q-HRs, being non-locus specific structures of genomes, are subject to additive selection, i.e., their total amount is controlled in the genome of any human population [Kurmanova 1991; Ibraimov 2015a; Ibraimov et al. 1986, 1990, 1991, 2000].

The data presented in Table 2.20.1.1 indicate that, at the level of individual human populations, each of the seven potentially Q-polymorphic autosomes contains a well-defined "share" of the total number of Q-variants in the genome, regardless of racial/ethnic characteristics [Ibraimov 1993, 2011]. Similar data was obtained through analyzing lowland (steppe Mongoloids of Central Asia), high altitude, and northern Mongoloid populations of Eurasia. If the frequencies of Q-variants are expressed in relative numbers (Q-HR frequency as a percentage of the overall number of chromosomal Q-HRs), then it becomes obvious that interpopulation heterogeneity is due to a proportional increase or decrease in the absolute number of Q-variants in all 12 potentially polymorphic loci of seven autosomes.

Unfortunately, it remains impossible to compare the data obtained in different laboratories for methodological reasons. However, when analyzing data from different laboratories separately, we are convinced that the distribution of Q-variants in seven Q-polymorphic autosomes is far from random and that each locus with Q-variants contains a certain "fraction" of the total Q-heterochromatic material in the genome of any human population, regardless of racial/ethnic origin and sex/age, as well as the characteristics of the ecological environment of its permanent habitat [Ibraimov 1993, 2010, 2011a; Ibraimov et al. 2013].

Therefore, there is reason to believe that the heterogeneity of human populations, in terms of the distribution of Q-variants, is not due to different degrees of polymorphism in certain loci in different samples, but due to a proportional increase or decrease in the content of Q-HRs simultaneously present in all potentially polymorphic loci of seven autosomes, which best reflects such a quantitative characteristic of chromosomal Q-HR variability, as the mean number of Q-HRs calculated per individual in a population [Ibraimov 1993; Ibraimov et al. 1986].

As has been known since the time of J. B. S. Haldane, the steady-state frequency of a gene in a population corresponds to certain equilibrium between the rate of mutation and the intensity of selection. We do not suggest either the selective mutability of 12 mutable loci, as having fluorescent allelomorphs, or the locus-specificity of the Q-variant function, which would determine the different selection intensity for these Q-polymorphic loci of seven autosomes.

It is also not possible to explain the differences in Q-HR frequencies in loci as due to the effects of genetic drift, which always occurs in limited populations; the similarity of distributions of relative frequencies of Q-variants in loci for very different populations would require an assumption that, in all these populations, drift was realized on one side, i.e., displaying an accumulated "brilliant" and having a fluorescence intensity at the 4th and 5th degrees, with alleles of the 3cen and 13p11 loci and similar alleles of all other loci eliminated.

Therefore, we attempt to explain differences in Q-HR frequency by loci, in terms of the statement on the adaptability of chromosomal Q-HRs formulated earlier [Ibraimov et al. 1982, 1986]. The adaptability of a genotype depends on the total number of Q-HRs in all 12 loci of the seven autosomes.

We understand that the model of additive selection, expressed in the above assumptions, can be overly simplistic when compared to real natural processes. As such, here we consider it to be a qualitative demonstration of the fundamental possibility of interpopulation maintaining differences in Q-HR frequencies without the assumption of the mutational or functional specificity of loci involved in additive selection.

We have previously divided these polymorphic sites of seven autosomes into Q-high-frequency (Q-HF) and Q-low-frequency (Q-LF) loci [Ibraimov et al. 1986]. Q-HF loci include those Q-polymorphic segments that contain more than 20 % of all Q-variants, while Q-LF loci include those segments in which the frequency of Q-variants does not exceed 10 % of the total number of chromosomal Q-HR in a population's genome. Having surveyed more than two dozen human populations, we have been unable to detect a single sample in which the frequency of Q-variants on Q-HF loci (3 cen and 13p11) was less than 20 %. According to our data, on all other Q-polymorphic loci the frequency of Q-variants has never exceeded 10 %, allowing their attribution to Q-LF loci [Ibraimov et al. 1986]. Therefore, if the frequencies of Q-variants are expressed in relative numbers, then it becomes obvious that interpopulation heterogeneity is due to a proportional increase or decrease in the absolute number of Q-variants in all 12 potentially polymorphic loci of seven autosomes.

It is obvious that, in different populations, this process is accompanied by a whole set of microevolutionary processes of different intensities (drift, correlated selection, etc.). Therefore, the similarity of the relative frequencies of Q-variants between populations is manifested in the fact that the Q-HF loci are the same (it is natural to assume they have an early evolutionary origin), as with the Q-LF ones. However, for different populations, the ranking of the latter in relative frequency terms is different.

We hoped that the effect of increasing differences in frequencies between the Q-HF and Q-LF loci could be observed in our material. In Fig. 2.20.1.1, the populations we studied are ordered by the mean number of Q-variants per individual [Kurmanova 1991].

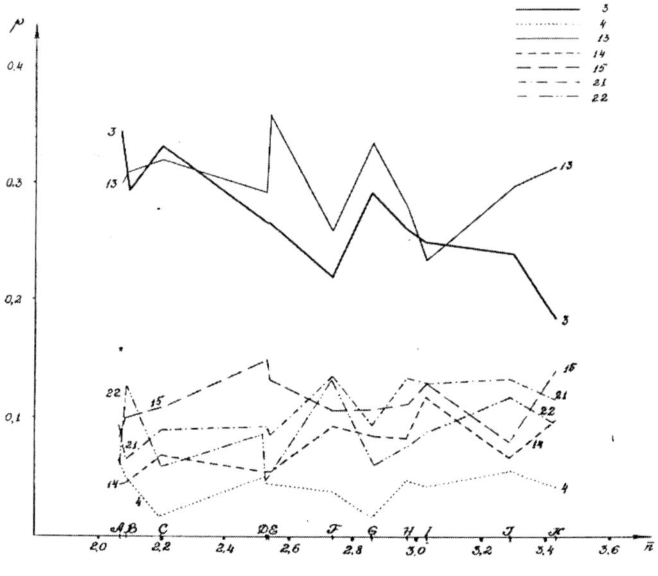

Fig. 2.20.1.1. The distribution of relative frequencies of Q-HRs across seven variable autosomes. A – Kyrgyz (Kyzyl-Dzhar, Pamir) *; B – Kyrgyz (Batken, Pamir) *; S – Chukchi **; D – Kyrgyz (Naryn, Tien Shan); E – Khakas **; F – Kyrgyz (Daraut-Korgon, Pamir) *; G – Russians ***; I – Turkmen ****; J – Kazakhs *; K – Chinese *[Ibraimov et al. 1982; ** Ibraimov and Mirrakhimov 1982a; *** Ibraimov and Mirrakhimov 1982b; **** Ibraimov and Mirrakhimov 1985].

For each of them, the relative frequencies of Q-variants on seven chromosomes are indicated (since this figure is not intended for rigorous quantitative assessment, we have combined the p11 and p13 loci of the acrocentric chromosomes). The number of populations analyzed was too small to statistically confirm the above assumption and we consider it sufficient to highlight some qualitative similarities in these graphs.

First, the assumption we made [Ibraimov 1993, 2011; Ibraimov et al. 1986] that the distribution of relative frequencies of Q-variants, on average, is the same for different populations has been confirmed. It can be seen from the figure that the relative frequencies of Q-variants for individual loci vary up to a change in rank from population to population, but it may be considered that the main reason for this variability is the impact of a purely selective effect. At the same time, the differences between high and low-frequency loci persist.

Secondly, the figure shows that the apparent dependence on the mean number of Q-variants per individual does not show the relative frequencies of Q-alleles on chromosomes 4, 13, 15, and 22.

At the same time, an increase is evident for chromosomes 14 and 21, and for chromosome 3 we see a decrease in relative frequency with an increase in the mean number of Q variants from 2.07 to 3.43. Since the decrease in the mean number of Q-variants is assumed to be the result of an increase in the intensity of additive selection (in the high altitudes and the Extreme North for populations on the left side of the scale), we can assume that this graph gives a quantitative confirmation of the assumption made by us to increase the differentiation between Q-high and Q-low-frequency loci with an increase in the selection coefficient as the number of Q-variants in the karyotype of an individual increases.

2.20.2. The mean number of chromosomal Q-HRs in the genome of human populations living in different climatic and geographical conditions

As can be seen in the previous section, human populations do not differ significantly in the pattern of the frequency distribution of Q-variants in 12 potentially polymorphic loci of seven autosomes. In this case, it may be expected that interpopulation differences are mainly due to differences in the total number of Q-HR variants in seven Q-polymorphic autosomes in the gene pool of one human population or another, which best reflects the quantitative characteristic of the variability of chromosomal Q-HRs as the mean number of Q-HRs per individual in a population [Ibraimov 1993; Ibraimov and Mirrakhimov 1985].

The value of the mean number of Q-HRs calculated per individual in a population, as an important quantitative characteristic of chromosomal Q-HR variability, has not yet been properly evaluated. Almost all population studies are limited, basically, by giving the numerical value of this characteristic without any discussion (see Section 2.13).

Lubs et al. [1977], who first discovered the existence of differences in the mean number of Q-variants, based on a comparative analysis of the variability of chromosomal Q-HRs, and samples of "white" (n = 205) and "black" (n = 210) children in the United States aged 7-8 years, limited themselves to a statement of the fact that interracial cytogenetic differences exist and did not offer any explanation or possible reason for the phenomenon discovered, implying that such a difference is due to the ancestor effect. A similar trend in the interpretation of data from comparative cytogenetic studies has been noted in other works [Müller et al. 1975; Al-Nassar et al. 1981; Nazarenko 1987].

Buckton et al. [1976] found a significant decrease in the mean number of Q-variants in one island sample in Scotland when compared to the mainland population. They limited themselves remarking that such a difference could be age-related rather than population-related, since the island sample was composed of people of 65 years and older and the mainland population sample was composed of newborns and 14-year-old boys and girls. Table 2.13.1 presents data on the mean number of Q-variants of those authors who have used similar criteria for recording Q-heterochromatic chromosome segments. On average, this value is 3.8.

We carried out a comparative study of the frequency and types of Q-HR variants on seven pairs of Q-polymorphic autosomes (3, 4, 13-15, 21, and 22) in three steppe (Kazakh, Chinese, Mongol) and three high altitude (Pamir and Tian-Shan) Mongoloid populations of Central Asia. The steppe Mongoloid populations showed statistically significant uniformity across all the studied quantitative characteristics of the variability of chromosomal Q-HRs. The same results were obtained when comparing the highlanders of the Pamirs (2 samples) and the Tien-Shan. However, a comparative analysis of the steppe and high altitude Mongoloids showed statistically significant heterogeneity in terms of the mean number of Q-HRs in the population with 2.5 for high altitude and 3.4 for steppe Mongoloids [Ibraimov et al. 1982].

We consider it logical that if chromosomal Q-HRs really have the selective value we assume, then a sample of people who showed good adaptive ability under the extreme conditions of high altitude should be investigated. For this reason, we studied a large group of mountaineers [Ibraimov et al. 1990]. During 1984-1986, we surveyed 385 mountaineers representing different national and ethnic groups. Of these, 277 had parents of Russian nationality, 63 had only one Russian parent, and the rest (46) represented one of more than 20 ethnic groups living in the USSR. Therefore, the first sample of mountaineers was called Russian, the second "half-Russian", and the third "mestizo". The age range of the athletes was 17 to 54 years. Of these, 282 were male and 103 were female.

At present, there is no unified international rule for assessing the qualifications of mountaineers. Our sample included: individuals who had been engaged in mountaineering for at least 3 years and climbed mountain peaks not lower than 4,500 m; highly skilled athletes who had conquered all 7,000 m peaks of the USSR (the so-called "snow leopards"); as well as members of the Soviet Himalayan expedition that reached the summit of Everest. At the same time, we simultaneously studied several high altitude permanent Kyrgyz settlements in the Eastern Pamirs (2,600-3,600 m above sea level as a "control") in order to compare them with mountaineers, as well as Russians living in Kyrgyzstan for comparison with the Russian athletes. The corresponding comparative analysis of all three samples (Russian, semi-Russian, and mestizo) of the mountaineers showed their statistically significant uniformity over all studied quantitative characteristics of the variability of chromosomal Q-HRs (tables 2.20.2.1-2.20.2.3). Table 2.20.2.1 shows the frequency of Q-HRs in all 12 polymorphic loci of seven autosomes among the mountaineers, taking into account their ethnic origin. Table 2.20.2.2 shows the results of the same comparative analysis by gender. Table 2.20.2.3 shows the frequency of Q-variants for the athletes belonging to different age groups. Based on these results, we combined them into a single group [Ibraimov et al. 1990].

Table 2.20.2.1. Q-HR frequencies in twelve polymorphic loci of seven autosomes in male and female mountaineers.

Location of Q-HRs	Males (n = 287)	Females (n = 103)	Hypergeometric test values
3 cen	179	67	061
4 cen	8	7	0.98
13 p11	134	52	0.70
13p13	12	2	0.23
14p11	17	8	0.80
14p13	16	4	0.34
15p11	19	5	0.34
15p13	18	3	0.14
21p11	16	5	0.49
21p13	26	3	*0.03**
22p11	16	2	0.10
22p13	6	3	0.80
Total Q-HRs	467	161	$\chi^2 = 12.44$ df = 7** P >0.05

* Statistically significance differences.

** Here and elsewhere combined classes preliminarily classified according to n_{iJ} values such that $n_{ij} \geq 5$.

Table 2.20.2.2. Q-HR frequencies in twelve polymorphic loci of seven autosomes in male and female mountaineers of different nationalities.

Location of Q-HRs	Russians (n = 277) 1	Half-castes (n = 63) 2	Mixed (n = 45) 3	Hypergeometric test values 1-22-31-3		
3 cen	183	37	26	0.25	0.53	0.26
4 cen	7	4	4	0.96	0.44	0.99
13p11	130	32	24	0.72	0.64	0.79
13p13	9	2	3	0.67	0.35	0.93
14p11	19	4	2	0.57	0.51	0.42
14p13	10	3	7	0.79	0.64	0.99
15p11	17	6	1	0.89	0.13	0.26
15p13	14	5	2	0.88	0.38	0.61
21p11	14	2	5	0.41	0.11	0.96
21p13	20	5	4	0.69	0.70	0.77
22p11	13	3	2	0.66	0.65	0.65
22p13	8	0	1	0.19	0.42	0.63
Total Q-HRs	444	103	81	$\chi2 = 1.59$; df = 6; $P > 0.95$;	$\chi2 = 1.44$; df = 3; $P > 0.50$;	$\chi2 = 13.3$; df = 5; $P < 0.01$

Table 2.20.2.3. Q-HR frequencies in twelve polymorphic loci of seven autosomes in mountaineers of different ages.

Location of Q-HRs	<19 yr (n = 22) 1	20-29 yr (n= 241) 2	30-39 yr (n = 87) 3	40+ yr (n = 35) 4	Hypergeometric test values					
					1-2	1-3	1-4	2-3	2-4	3-4
3 cen	13	159	54	20	0.81	0.69	0.55	0.78	0.89	0.76
4 cen	1	8	4	2	0.83	0.74	0.67	0.81	0.88	0.77
13p11	7	124	42	13	0.06	0.13	0.45	0.35	0.80	0.18
13p13	0	13	0	1	0.31	>0.99	0.61	*0.02*	0.45	0.29
14p11	0	20	4	1	0.16	0.40	0.61	0.19	0.22	0.55
14p13	1	11	6	2	0.73	0.57	0.67	0.87	0.78	0.59
15p11	1	16	4	3	0.57	0.74	0.50	0.35	0.79	0.90
15p13	2	13	3	3	0.88	0.95	0.71	0.35	0.87	0.23
21p11	1	12	5	3	0.70	0.65	0.50	0.72	0.89	0.84
21p13	1	20	5	3	0.46	0.65	0.50	0.31	0.67	0.84
22p11	2	10	4	2	0.93	0.90	0.50	0.70	0.82	0.77
22p13	1	7	1	0	0.86	0.96	0.39	0.33	0.38	0.71
Total Q-HRs	30	413	132	53	$\chi^2 = 1.52$ df = 2 P >0.70	$\chi^2 = 0.96$ df = 2 P >0.50	$\chi^2 = 0.36$ df = 2 P >0.80	$\chi^2 = 2.90$ df = 5 P >0.70	$\chi^2 = 2.45$ df = 3 P >0.30	$\chi^2 = 2.66$ df = 3 P >0.20

Table 2.20.2.4. The distribution and mean numbers of Q-HRs per individual in mountaineers.

Number of Q-HRs	Males (n = 282)	Females (n = 103)	Russians (n = 277)	Half-castes (n = 63)	Mixed Group (n = 45)	<19 yr (n = 22)	20-29 yr (n = 241)	30-39 yr (n = 87)	40+ yr (n = 35)
0	41	16	46	6	5	5	27	19	6
1	84	32	81	22	13	6	77	21	12
2	101	40	100	25	16	9	91	31	10
3	44	12	39	9	8	2	32	15	7
4	11	2	9	1	3	0	12	1	0
5	1	1	2	0	0	0	2	0	0
Total Q-HRs	467	161	444	103	81	30	413	132	53

$\chi^2 = 1.43$; df = 3; $p > 0.50$;
$\chi^2 = 3.62$; df = 6; $P > 0.70$;
$\chi^2 = 2.24$; df = 3; $P > 0.50$;

Mean number of Q-HRs	1.66	1.56	1.60	1.63	1.80	1.36	1.71	1.52	1.51

Table 2.20.2.5. Q-HR frequencies in twelve polymorphic loci of seven autosomes in mountaineers and Russian control.

Location of Q-HRs	Mountaineers (n = 277)	Russians (n = 200)	Hypergeometric Test values
3 cen	183	168	0.99
4 cen	7	10	0.12
13p11	130	178	*<0.01*
13p13	9	14	0.05
14p11	19	29	*<0.01*
14p13	10	20	*<0.01*
15p11	17	53	*<0.01*
15p13	14	10	0.58
21p11	14	18	0.07
21p13	20	37	*<0.01*
22p11	13	20	*0.02*
22p13	8	15	*0.02*

Total Q-HRs 444 572

$\chi^2 = 27.84$; df = 11; P <0.001

Table 2.20.2.6. The distribution and mean numbers of Q-HRs per individual in mountaineers and Russian control.

Number of Q-HRs	Mountaineers (n = 277)	Russians (n = 200)
0	46	9
1	81	29
2	100	49
3	39	54
4	9	34
5	2	24
6		4
7		1

Total Q-HRs 444 572

$\chi^2 = 97.05$; df = 4; P <0.001;

Mean number of Q-HRs 1.60 ± 0.06; 2.86 ± 0.10

t = 10.40; df = 410; P <0.001;

Table 2.20.2.7. Q-HR frequencies in twelve polymorphic loci of seven autosomes in mountaineers and Kyrgyz control.

Location of Q-HRs	Kyrgyz (n = 284)	Mountaineers (n = 385)	Hypergeometric test values
3 cen	204	246	0.94
4 cen	38	18	*<0.01*
13p11	157	183	0.95
13p13	21	14	0.03
14p11	12	25	0.14
14p13	14	20	0.51
15p11	27	24	0.08
15p13	22	21	0.15
21p11	36	21	*<0.01*
21p13	20	29	0.47
22p11	29	18	*<0.01*
22p13	9	9	0.81
Total	589	628	

$\chi^2 = 27.23$; df = 11; P <0.01

Table 2.20.2.8. The distribution and mean numbers of Q-HRs per individual in mountaineers and Kyrgyz control.

Number of Q-HRs	Kyrgyz (n = 284)	Mountaineers (n = 385)
0	32	57
1	67	116
2	87	141
3	56	56
4	30	13
5	12	2
Total Q-HRs	589	628

$\chi^2 = 31.18$; df = 4; p <0.001

Mean number of Q-HRs 2.07 ± 0.08; 1.63 ± 0.05

t = 4.716; df = 531; P <0.001

However, as can be seen in tables 2.20.2.4 and 2.20.2.5 there is a significant difference between mountaineers of Russian nationality and Russians from the control group in terms of the total quantitative content of chromosomal Q-HRs—in the genome of athletes, the number of Q-HRs is significantly less than in the control. A similar comparative analysis was conducted comparing highlanders and mountaineers (tables 2.20.2.6 and 2.20.2.7). Contrary to our expectations, the Q-HR content in the genome of the mountaineers was statistically significantly lower than that of the indigenous inhabitants of the high altitude Pamir Mountains. This difference may be due to a number of factors: 1) although the highlanders live permanently at high altitudes, they rarely go past 4,000 m above sea level, as they are mainly engaged in livestock animal husbandry; 2) muscular and mental stress may play a part and the magnitude and intensity of physical exertion experienced by mountaineers while climbing the mountain peaks is rare in the measured life of high altitude inhabitants; 3) the Pamir highlanders have lived in high altitude conditions for many centuries and have rich behavioral and ecological experiences that affect their ability to acclimatize and reduce the effects of harmful physical factors of great heights. Since the mountaineers we surveyed were all native lowlanders, the demands that the high altitude environment made on their bodies were probably very high. Many years of experience in communicating with and observing athletes made it possible to conclude that, to become a mountaineer, a love of mountain peaks alone is not enough, one must also have the appropriate natural attributes. The fact is that the majority of young people who visited the mountaineering sections, despite their success in climbing and many years of training, did not breach the altitude of 4,500 meters above sea level and left mountaineering [Ibraimov 2019c; Ibraimov et al. 1990].

The results of a comparative analysis of the indigenous inhabitants of high altitudes and mountaineers prompted us to investigate the aborigines of the Extreme North of Siberia and foreign individuals to try to find out whether low temperature, as one of the environmental factors that limits human life, affects the pattern of chromosomal Q-HR variability.

The Eurasian circumpolar belt is inhabited by people who differ significantly from each other in ethnic composition, language, culture, and way of life. In recent decades, this region has been intensively settled and modified by Russians and other ethnic groups after the discovery of vast oil and gas reserves. Therefore, we chose Russians who had been engaged in physical labor (drillers) for many years under the open sky of the Yamal Peninsula of Western Siberia, and their children who were born there. As a control, we sampled students from the indigenous populations studying in upper-school classes: Selkup (v. Krasnoselsk), Khanty (v. Muzhi), and Yakuts (city of Salekhard).

Table 2.20.2.9 shows the distribution and mean number of Q-HRs per individual in the population. As can be seen in this table, there are no significant differences between the three samples taken from Eastern

Siberian natives in terms of the quantitative content of Q-HRs in the genome, despite being representatives of very distant ethnic groups of Mongoloids. Comparative analysis of the remaining quantitative characteristics of chromosomal Q-variants also did not show the presence of any significant differences between them, allowing us to combine them into one group [Ibraimov et al. 1991].

Table 2.20.2.9. The distribution and mean numbers of Q-HRs per individual among Selkups, Khants, and Yakuts.

Number of Q-HRs	Selkups (n = 90) I	Khants (n = 54) II	Yakuts (n = 127 III
0	13	6	17
1	19	17	32
2	39	20	43
3	11	6	23
4	5	4	10
5	3	1	2
Total Q-HRs	165	96	237
	$\chi_{I,II}^2 = 2.10$; df = 4; P >0.70;	$\chi^2_{II,III} = 1.98$; df = 4; P >0.70;	$\chi^2_{I,III} = 2.85$; df = 4; P >0.50
Mean number of Q-HRs	1.83 ± 0.12;	1.78 ± 0.16;	1.87 ± 0.11;
	$t_{I,II} = 0.27$; df = 142; P >0.50;	$t_{II,III} = 0.17$; df = 179; P >0.50;	$t_{I,III} = 0.02$; df = 215; P >0.50

Table 2.20.2.10 shows the distribution of Q-HRs in the genomes of indigenous people sampled, their children, workers, and the control group made up of Russian nationals. Indigenous people and workers displayed a similar Q-HR content in their genomes (pairs I and III). However, there were statistically significant differences between workers, schoolchildren, and the control (couples II, III and III, IV), as well as between schoolchildren and the control of Russian nationals (couples II, IV).

Table 2.20.2.10. The distribution and mean numbers of Q-HRs per individual among natives, Russian children, oil workers, and the control group.

Number of Q-HRs	Natives (n = 271) I	Russian children (n = 113) II	Oil workers (n = 43) III	Controls (n = 200) IV
0	36	9	4	9
1	68	21	13	25
2	102	27	19	49
3	40	39	5	54
4	19	12	2	34
5	6	3		24
6		2		4
7				1
Total Q-HRs	498	267	74	57

$\chi^2_{I,II} = 25.6$; df = 5; $P < 0.001$; $\chi^2_{I,III} = 2.3$; df = 3; $P > 0.50$; $\chi^2_{I,IV} = 66.2$; df = 5; $P < 0.001$; $\chi^2_{II,III} = 13.3$; df = 3; $P < 0.01$; $\chi^2_{II,IV} = 13.5$; df = 5; $P < 0.05$; $\chi^2_{III,IV} = 20.8$; df = 3; $P < 0.001$

Mean number of Q-HRs: 1.84 ± 0.07; 2.36 ± 0.12; 1.72 ± 0.15; 2.86 ± 0.10

$t_{I,II} = 3.82$; df = 382; $P < 0.001$; $t_{I,III} = 0.63$; df = 312; $P > 0.50$; $t_{I,IV} = 8.2$; df = 395; $P < 0.001$; $t_{II,III} = 2.92$; df = 154; $P < 0.01$; $t_{II,IV} = 3.07$; df = 311; $P < 0.01$; $t_{III,IV} = 6.40$; df = 89; $P < 0.001$

Interestingly, when analyzing the mean numbers of Q-HRs per individual in these populations, the highest value was found in the control sample. Slightly low values for the mean numbers of Q-HRs were found in children, while the lowest was shown by oil workers. Curiously, the number of Q-HRs in the sample of Russian workers was slightly lower than that of indigenous people.

Above, we showed that the mean numbers of Q-HRs of chromosomes in the genome of mountaineers was lower than that of the highlanders of the Pamir Mountains. The results of a comparative study of foreigners and indigenous people of Eastern Siberia describe a similar trend in the conditions of the Extreme North [Ibraimov et al. 1991]. Thus, the quantitative content of Q-HRs in the genome of Russian oil workers, who were living and working in the conditions of the circumpolar belt of Eastern Siberia, was close to those of indigenous people, but differed from the control. It is noteworthy that a difference was found between Russian schoolchildren, who, in essence, are migrants of the first or second generation, and schoolchildren of Russian nationality living in Kyrgyzstan (control group). Our considerations of possible reasons for the heterogeneity of human populations in terms of the quantitative content of chromosomal Q-HRs in the genome living in different natural conditions, including those at high altitude and in the Extreme North, are outlined in Part III.

The data presented above show that human populations do not differ fundamentally from each other according to the distribution frequency of Q-variants and interpopulation differences are mainly due to a proportional increase or decrease in the content of Q-HRs for all Q-polymorphic autosomes (see Section 2.20.1). In this case, we may expect that, with an increase or decrease in the value of the mean number of Q-variants in a population, there will be a proportional increase or decrease in the absolute number of Q-HRs in all seven Q-polymorphic autosomes.

To test the validity of this assumption, we used a simple method where all the samples we studied were arranged in rows according to the mean numbers of Q-variants per individual in these populations and the frequencies of Q-variants of seven Q-polymorphic autosomes in the studied populations showed an increase (Table 2.20.2.11).

Table 2.20.2.11. Distribution of populations according to the mean number of Q-HRs per individual and the frequency of Q-HRs on seven Q-polymorphic autosomes.

P	n	*m*	3[a]	4[a]	13[a]	14[a]	15[a]	21[a]	22[a]
I	385	1.63	I 131.9[b]	I 1.9	I 26.0	I 5.8	I 5.8	I 6.5	I 3.5
II	520	2.01	III 33.6	IV 2.5	II 31.9	II 6.0	II 8.3	II 11.7	II 4.6
III	603	2.32	II 34.4	II 3.1	III 33.6	III 6.5	III 13.0	III 12.0	III 11.6
IV	297	3.49	IV 37.5	III 6.4	IV 49.1	IV 14.8	IV 22.2	IV 24.1	IV 18.0
V	327	4.66	V 64.4	V 8.2	V 77.6	V 17.3	V 23.9	V 24.1	V 22.8

P – populations; n – sample size; *m* – mean number of Q-HRs per individual; a – number of autosome; b – Q-HR frequency from the number of chromosomes analyzed; I – Mountaineers; II – northern Mongoloids; III – highland Mongoloids; IV – steppe Mongoloids; V – lowland Negroids.

From analyzing this table, it is clear that the distribution frequencies of Q-variants in seven Q-polymorphic autosomes in the populations, generally, correspond to their distribution as the mean number of Q-variants per individual in the study populations. For example, the northern and high altitude populations, which have the lowest values of mean number of Q-variants, mainly occupy the upper lines, and the lowland natives of subequatorial Africa, distinguished by the high mean number values of Q-HRs per individual in the populations, are assigned the lowest lines of the table. This trend is most evident in Q-high-frequency chromosomes.

Thus, it may be argued that interpopulation differences arise not due to the different degree of polymorphism of one or other locus with Q-variants, but due to a proportional increase or decrease in the frequency of Q-HRs in all Q-polymorphic loci of the seven autosomes. In this case, the most meaningful interpopulation differences can be expressed by the mean number of Q-variants calculated per individual in a population [Ibraimov et al. 1986, 1990].

Analysis of the nature of interpopulation differences in terms of the quantitative content of chromosomal Q-HRs, expressed as the mean number of Q-variants in the sample, has shown the existence of another, previously unnoticed pattern in the distribution of Q-variants in human populations. As noted above, individuals with different numbers of Q-HRs of chromosomes are found in various populations. Table 2.14.1 presents data on the distribution of Q-variants in the samples, along with the mean number values of Q-variants per individual in the study populations.

2.20.3. Chromosomal Q-HR variability and sex in human populations

The literature shows the absence of statistically significant differences between the sexes in the frequency of Q-variants in 12 polymorphic loci of seven autosomes [Geraedts and Pearson 1974; Lin et al. 1976; Buckton et al. 1976; Schwinger and Wehner 1977; Yamada and Hasegawa 1978; Ibraimov et al. 1982 a, b; Ibraimov and Mirrakhimov 1982 a, b, c, 1985].

Unfortunately, not all authors present primary quantitative data on the variants of chromosomal Q-HRs found in the samples they studied, limiting themselves only to statements of fact on the absence of sex differences. This makes it difficult to carry out any additional statistical analysis to find out whether there are really no differences by sex in the quantitative characteristics of chromosomal Q-variants. When comparing the data obtained by us in the analysis of the variability of Q-HRs on autosomes in a population, we have repeatedly drawn attention to this issue; in almost all cases, we have seen a tendency towards an increase in the mean number of Q-HRs in women when compared to men. However, in no sample were sex differences statistically significant [Ibraimov 1983; Ibraimov and Mirrakhimov 1985; Ibraimov et al. 1986, 1991]. Table 2.20.3.1 shows the results of a comparative analysis of the samples and we were able to examine a sufficiently large number of men and women from the same population.

Table 2.20.3.1. The mean numbers of chromosomal Q-HRs in males and females in the same population.

Population	Sex	n	*m*	t	df	P
Kyrgyz (Batken)	M F	43 52	2.05±0.05 2.12±0.18	0.22	92	>0.50
Kyrgyz (Naryn)	M F	42 71	2.45±0.20 2.59±0.17	0.53	111	>0.50
Kyrgyz (Alay)	M F	42 70	2.50±0.24 2.89±0.20	1.27	110	>0.20
Kyrgyz (Kyzyl-Djar)	M F	98 100	1.92±0.12 2.17±0.13	1.38	196	>0.10
Kyrgyz (Sary-Tash)	M F	31 16	2.00±0.20 2.00±0.37	0.00	45	>0.99
Kyrgyz (Murgab)	M F	20 19	2.10±0.35 2.52±0.34	0.87	37	>0.20
Chinese (North China)	M F	52 72	3.31±0.20 2.74±0.20	1.47	122	>0.10
Kazakhs (Kazakhstan)	M F	40 61	3.55±0.23 3.57±0.24	0.06	100	>0.50
Khakass (Russia)	M F	49 71	2.43±0.16 2.61±0.15	0.74	118	>0.20
Chukchy (Russia)	M F	55 77	2.09±0.17 2.27±0.13	0.86	131	>0.20
Yakuts (Russia)	M F	46 81	1.70±0.14 1.96±0.15	1.40	123	>0.10
Selkups (Russia)	M F	33 57	1.64±0.20 1.95±0.16	1.20	88	>0.20
Khants (Russia)	M F	22 32	1.45±0.22 2.00±0.21	1.77	52	>0.05
Russians (Russia)	M F	106 94	2.81±0.14 2.91±0.15	0.40	198	>0.50
Turkmen (Turkmenia)	M F	45 71	3.00±0.24 3.04±0.20	0.13	114	>0.50

n – sample size; m – mean number of Q-HRs; t, df, and P – statistics.

As can be seen in this table, there are no statistically significant differences between the sexes in terms of the mean number of chromosomal Q-HR variants. However, in all the samples examined a tendency towards a greater mean number of Q-variants per individual in women, compared to men, was seen.

The quantitative content of chromosomal Q-HRs in the genome of a population can be estimated, in addition to their frequency on Q-polymorphic loci of seven autosomes, using the mean number, size, and fluorescence intensity of Q-HR variants of chromosomes (Paris Conference 1971, Suppl. 1975). Using identical methods for evaluating all the quantitative characteristics of chromosomal Q-HRs, we examined infants and adults (students of KazNMU in Almaty) to find out if there were any differences by sex at the population level in two ethnic groups of Kazakhstan (Kazakhs and Russians). Table 2.20.3.2 shows the distribution and mean numbers of Q-HRs of autosomes in newborns and adults of Kazakh nationality by gender.

Table 2.20.3.2. The distribution and mean numbers of Q-HRs on autosomes in males and females in newborns and adults (18-25) of Kazakh nationality.

Number of Q-HRs	Newborns Boys (n = 207) I	Newborns Girls (n = 182) II	Males 18-25 years (n = 49) III	Females 18-25 years (n = 190) IV
0	3	1		
1	5	4	9	7
2	39	21	12	24
3	47	38	13	45
4	51	46	11	49
5	37	39	4	40
6	18	20		17
7	7	13		7
Total number of Q-HRs	770	750	136	745
Mean number of Q-HRs	3.72 ± 0.102	4.12 ± 0.111	2.78 ± 0.176	3.92 ± 0.104
Statistics	t I, II = 2.649; df = 387; P = 0.008* t II, III = 5.775; df = 229; P = <0.001* t III, IV = 5.119; df = 237; P = <0.001* t I, III = 4.137; df = 254; P = <0.001* t II, IV = 1.313; df = 370; P = 0.190			

* These differences are statistically significant.

As can be seen in this table, in each case, the female sample was characterized by high mean number values and a wider range of variability of chromosomal Q-HRs compared to males. These differences are statistically significant. The same patterns were found in the Russian samples [Ibraimov 2014].

Additionally, it is possible to estimate the quantitative content of chromosomal Q-HRs in the human genome by size and fluorescence intensity using the five-degree system recommended by the Paris Conference [Paris Conference 1971, 1975]. When evaluating the size of chromosomal Q-HRs, we followed the recommendations of Yamada and Hasegawa [1978], according to which the sizes of Q-heterochromatic bands are conventionally divided into five degrees and comparing them to the size of the short arm of chromosome 18. At the same time, the fluorescence intensity of the Q-HR variants was estimated using the five-degree system of the Paris Conference.

The abbrevation "QFQ" means "Q-bands by fluorescence using quinacrine" [Paris Conference 1970, 1975]. For example, the abbreviation "QFQ 15" means that this Q-HR is of average size and has very bright fluorescence (5th degree of fluorescence intensity).

Among the individuals studied, the values of QFQ types ranged from 15 to 112. Here, QFQ 15 means that the karyotype of this individual contains only one medium-size Q-HR with the 5th degree of fluorescence, while, for example, QFQ 98 means that in the karyotype of this individual there are seven Q-HRs of medium size achieving the 5th degree of fluorescence intensity. For the statistical analysis of QFQ types, they are conditionally divided into 7 groups: group 1 (indicators QFQ 14, 15); group 2 (indicators QFQ 28, 29); group 3 (indicators QFQ from 40 to 45); group 4 (indicators QFQ from 53 to 58); group 5 (indicators QFQ from 67 to 73); group 6 (indicators QFQ from 80 to 86); and group 7 (indicators QFQ from 98 to 112). In other words, the first group includes individuals, in the karyotype of which there is only 1 Q-HR; the second group includes those with 2 Q-HRs; and the remaining groups represent individuals with 3 or more Q-HRs at the 5th degree of fluorescence intensity (Table 2.20.3.3).

Table 2.20.3.3. Types of QFQ autosomes in newborns and adults of Kazakh nationality by sex (Almaty).

Types of QFQ	Newborns		Males 18-25 years (n = 43) III	Females 18-25 years (n = 120) IV
	Boys (n = 180) I	Girls (n = 159) II		
0	3	2		
1	6	4	8	7
2	34	15	7	14
3	40	33	13	28
4	42	36	12	30
5	36	40	2	26
6	14	14	1	11
7	5	13		4
8		4		
Total QFQ	661	668	125	463
Mean number of QFQ	3.67 ± 0.110	4.20 ± 0.668	2.91 ± 0.193	3.83 ±0.132
Statistics	$t_{I, II}$ = 3.164; df = 337; P = 0.002* $t_{III, IV}$ = 3.807; df = 161; P <0.001*			

The question arises, what is the reason for sex differences at the human population level? We believe the reason for this difference is the presence of a Y chromosome in the male karyotype. As is well known, in the human population males differ in terms of Y chromosome length (see Section 2.18). We compared male individuals taken from the same population, who had Y chromosomes of different sizes. Table 2.20.3.4 shows the distribution and mean numbers of Q-HRs by autosomes in three samples of male individuals (Kazakh newborns), taking into account the size of their Y chromosome. The same data was obtained for all other racial/ethnic groups of Eurasian and African descent studied by us [Ibraimov 2014; Ibraimov et al. 2000].

Table 2.20.3.4. Distribution and mean numbers of autosomal Q-HRs in Kazakh newborn males with Y chromosomes of various sizes.

Number of Q-HRs	Large $Y \geq F$ (n = 53) I	Medium $F > Y > G$ (n = 102) II	Small $Y \leq G$ (n = 32) III
0	3		
1	5	1	
2	21	12	2
3	11	26	1
4	6	28	10
5	7	22	13
6		10	4
7		3	2
Total number of Q-HRs	139	406	150
Mean number of Q-HRs	2.62 ± 0.185	3.98 ± 0.129	4.69 ± 0.203
Statistics	t I, II = 6.077; df = 153; P = <0.001* t II,III = 2.748; df = 132; P = 0.007* t I,III = 7.223; df = 83; P = <0.001*		

* These differences are statistically significant.

These data give us reason to believe that: a) Y chromosome Q-heterochromatin, being the largest in the human karyotype, somehow "restricts" the total number of Q-HRs on autosomes in the male genome; 2) human chromosomal Q-HRs, apparently, are very close, if not identical, in nature, regardless of their localization in the karyotype.

Why was it so difficult to determine the existence of statistically significant differences between the sexes according to the quantitative content of chromosomal Q-HRs at the human population level? We believe that there are several reasons: 1) until now, all authors have compared male and female samples taken from the same population only by Q-HR frequencies on 12 Q-polymorphic loci of autosomes, leaving aside other quantitative characteristics of chromosomal Q-polymorphisms, such as the distribution of numbers, sizes, and fluorescence intensities of Q-HRs in a population; 2) in such comparative analyses, the age of the studied individuals was not taken into account. Our experience has shown that statistically significant sex differences are found only when samples of individuals from the same age sample are compared [Ibraimov 2014]; 3) the extremely wide polymorphism of Y chromosome sizes in the study populations was not taken into account. It was not considered possible to define a close quantitative relationship between Q-HRs localized on different chromosomes in the human karyotype in the sense that Y chromosome Q-heterochromatin somehow affects, or more precisely limits, the number of Q-HRs on the autosomes. In other words, no one had specifically tried to find out whether the size of the Q-heterochromatin segment of the Y chromosome affected the number of Q-HRs on autosomes in the human genome in males.

It is well known that the Q-heterochromatin of the long arm of the Y chromosome is of medium size, on average being two times larger than the Q-HRs on any of the seven Q-polymorphic autosomes in the human karyotype. Based on the above data, it may be assumed that the Q-HR of the Y chromosome, being the largest Q-heterochromatin material in human karyotypes, somehow limits ("restricts") the total amount of Q-HRs on autosomes in men.

It seems highly probable that such an explanation for the number of Q-HRs in autosomes in women compared to men appears to be related to some evolutionarily fixed mechanism that restricts the total amount ("dose") of Q-heterochromatin material in the human genome acceptable for a given population. At the population level, such a restriction may occur in women due to the absence in their karyotype of a chromosome capable of carrying Q-heterochromatic material comparable in size to the Q-HR on the Y chromosome in men. Such a "dose" limitation of Q-heterochromatin seems to apply to the male sex, since males with large Y chromosomes have fewer Q-HRs on autosomes and *vice versa* [Ibraimov et al. 2000]. Indeed, there are 25 loci in the human karyotype (3 cen, 4 cen, 13 p11, 13 p13, 14 p11, 14 p13, 15 p11, 15 p13, 21 p11, 21 p13, 22 p11, 22, p13, and Yq12) where Q-heterochromatin can potentially be detected. However, no one has yet been able to detect as many Q-heterochromatin loci on one karyotype; usually the number of Q-HRs in the human karyotype ranges from 0 to 10 [Yamada and Hasegawa 1978; Al-Nassar et al. 1981; Ibraimov and Mirrakhimov 1985; Ibraimov 2010].

Detection of the existence of significant differences by sex may make it necessary to revise the literature on the mean numbers of chromosomal Q-HRs per individual in a population. We give only one example taken from our research. In 1982, we examined a sample of young men and women of Kazakh nationality in a region of southern Kazakhstan for polymorphism of chromosomal Q-HRs [Ibraimov at al. 1982]. After 30 years, we were able to examine boys and girls from the same age and sex group in the city of Almaty [Ibraimov et al. 2013]. These two Kazakh samples do not differ from each other in the quantitative characteristics of chromosomal Q-polymorphism. For example, the mean number values of Q-HRs per individual in both populations were almost the same, with an average of 3.56 ± 0.168. We suggested at the time that this number probably reflected the quantitative content of Q-heterochromatin in the genome of southern Kazakhs. Now, after studying Kazakhs of different age groups, we have to admit that our conclusion was not correct, since the mean number values of Q-HRs per individual in samples of newborns, young, and elderly Kazakhs were 4.20 ± 0.668, 3.56 ± 0.168, and 2.55 ± 0.213, respectively [Ibraimov et al. 2013, 2014a]. It follows that when presenting and discussing data on the quantitative content of chromosomal Q-HRs in a given population, we must indicate the values of the mean number of Q-HRs separately by sex with an indication of age (see below).

We believe that, if the tendency for a greater mean number of Q-variants per individual in women really occurs at the level of human populations, then this should not be particularly surprising. For all researchers, speaking of the absence of sex differences in chromosomal Q-polymorphisms relates only to the frequency of Q-variants in 12 polymorphic loci of seven autosomes, leaving aside the extremely wide variability of Q-HR sizes and fluorescence intensities in chromosomes in human populations. As far as we know, no one has yet attempted to assess the possible "contribution" to the common pool of Q-HRs in the population of the Y chromosome of men, which, as is well known, has the largest block of Q-heterochromatic in the human karyotype. The absence of such attempts is, of course, quite understandable at this stage, since there are currently no methods for accurately determining the quantitative content of Q-heterochromatin material in chromosomes in the human genome (see Section 2.3.11). Such attempts have been made only with respect to the C-heterochromatic regions of chromosomes and then only in three pairs of autosomes: 1, 9, and 16 [Baliček et al. 1978; Podugolnikova et al. 1979; Erdtmann et al. 1987].

It seems to us that the following explanation of the tendency for a greater size in the mean number of Q-variants per individual in woman is legitimate. There is likely some evolutionarily fixed mechanism that "compensates" for the difference in the dose of Q-heterochromatic material in the genome of women at the level of individual human populations, resulting from the absence in their karyotype of a chromosome that carries an equally large Q-heterochromatin segment, like the Y chromosome, at the expense of the autosomes.

Nazarenko [1987], using data on the variability of Y chromosome length in northern Khanty (Western Siberia), hypothesized the existence of "reciprocal variability" of the content of Q-heterochromatin in the human genome, expressed in the fact that the frequency of brightly fluorescent Q-heterochromatin segments along the locus is *3 cen* in individuals with a subtotal deletion of heterochromatin of the Y chromosome are significantly lower than those in the control group of male carriers of the non-deleted Y chromosome. The author believes that "this result probably indicates the possibility of reciprocal variability in the content of some heterochromatin fractions, as a result of which their level in the human genome is maintained within certain optimal limits".

We assumed that corresponding tendencies could also be revealed when comparing individuals with Y chromosomes of different sizes. In particular, it was expected that individuals with large Y chromosomes would have lower, and with small Y chromosomes, higher Q-variant frequencies in the *3 cen* locus, compared to individuals with medium Y chromosomes. As described in the next section, no significant differences between the groups were found for any of these loci and thus we cannot confirm the data of Nazarenko in our own material.

However, the very fact that we have detected a decrease in Q-allele frequencies in the sample of male carriers of the Y chromosome, almost completely devoid of Q-heterochromatin material, from our point of view, can be explained in terms of the hypothesis on the possible selective value of the total amount of Q-HRs in an individual genome [Ibraimov et al. 1982, 1986, 1990]. It is possible that the differences seen between the group of male *del* Y carriers and the control on the frequency of Q-HRs localized at 3 *cen* were found, in this case, only at this locus, due to the fact that it is among the most Q-high-frequency ones. In this case, with an increase in the sample size (in Nazarenko only 57 men with *del* Y were examined), a similar increase in the frequencies of Q-HRs can be detected in other Q-polymorphic loci.

It appears that the wide quantitative variability of human chromosomal Q-HRs at the population level is subject to certain regularities, which are available to quantitative analysis. Above, I have repeatedly shown that populations with a low mean value of Q-variants in the genome, as a rule, have a narrow range of variability in the distribution of Q-variants, and *vice versa* [see also Ibraimov and Mirrakhimov 1985; Ibraimov et al. 1982, 1986, 1990]. Therefore, it seems to us that there is a close relationship between the mean numbers of Q-variants per individual, the distribution and frequencies of Q-variants on autosomes, and the size of Y chromosome Q-heterochromatin at the population level.

2.20.4. Chromosomal Q-HR variability in individuals of various ages in human populations

Individuals in a population differ in terms of the number, location, size, and fluorescence intensity of chromosomal Q-HRs. Some human populations also differ significantly from each other [Geraedts and Pearson 1974; Müller et al. 1975; Buckton et al. 1976; Lubs et al. 1977; Yamada and Hasegawa 1978; Al-Nassar et al. 1981; Ibraimov and Mirrakhimov 1982a, b, c; Ibraimov et al. 1982, 1986, 1990, 1991, 2000, 2013; Stanyon et al. 1988; Kalz et al. 2005; Decsey et al. 2006]. At the same time, it is not known whether individuals from different age groups differ in the number, location, size, and fluorescence intensity of chromosomal Q-HRs. This fact is very important for understanding the possible biological role of chromosomal Q-HRs in human life. Using identical methods for estimating the number of chromosomal Q-HRs, we have studied samples of newborns, secondary school students, students of higher educational institutions, drivers and mechanics of automobile enterprises, elderly residents of the Bishkek nursing home, as well as newborns, students, and residents of the Almaty (Kazakhstan) nursing home, representing three ethnic groups: Kyrgyz , Kazakhs, and Russians. It appears that the greatest quantity of chromosomal Q-HRs are found in the genomes of newborns and the smallest quantity in the genomes of the elderly (60 years and older), regardless of the racial and ethnic characteristics of the individuals examined. Tables 2.20.4.1-2.20.4.2 show the distribution and mean numbers of chromosomal Q-HR in individuals belonging to five age groups of Kyrgyz and Russian nationality.

Table 2.20.4.1. The distribution and mean numbers of chromosomal Q-HRs per individual in Kyrgyz samples (Bishkek).

Number of Q-HRs	Ages				
	Newborns (n = 145) I	17-19 years (n = 317) II	20-39 years (n = 112) III	40-59 years (n = 67) IV	>60 years (n = 23) V
0	4	21	7	8	3
1	19	70	20	13	4
2	23	105	41	20	11
3	38	71	19	14	2
4	37	39	16	10	3
5	16	9	7	2	
6	5	2	2		
7	3				
Total Q-HRs	458	706	270	145	44
Mean Number of Q-HRs	3.16 ± 0.13	2.23 ± 0.07	2.41 ± 0.13	2.16 ± 0.16	1.91 ± 0.24
	$t_{I, II} = 6.0$ df = 232 $P < 0.000$	$t_{I, III} = 4.01$ df = 255 $P < 0.000$	$_{I, IV} = 4.54$ df = 210 $P < 0.000$	$t_{I, V} = 4.58$ df = 38 $P < 0.000$	$t_{II, III} = 1.29$ df = 427 $P > 0.20$
	$t_{II, IV} = 0.42$ df = 382 $P > 0.600$	$t_{II, V} = 1.21$ df = 338 $P < 0.200$	$t_{III, IV} = 1.20$ df = 177 $P > 0.200$	$t_{III, V} = 1.64$ df = 133 $P < 0.100$	$t_{IV, V} = 0.81$ df = 88 $P < 0.300$

Table 2.20.4.2. The distribution and mean numbers of chromosomal Q-HRs per individual in Russian samples (Bishkek).

Number of Q-HRs	Ages				
	Newborns (n = 37) I	17-19 years (n = 67) II	20-39 years (n = 115) III	40-59 years (n = 230) IV	>60 years (n = 211) V
0	0	4	6	14	26
1	3	9	18	50	51
2	7	21	39	74	81
3	5	19	29	58	35
4	12	9	17	27	13
5	7	4	6	6	4
6	3	1		1	1
Total Q-HRs	133	170	281	516	396
Mean Number of Q-HRs	3.59 ± 0.23	2.54 ± 0.16	2.44 ± 0.11	2.24 ± 0.08	1.88 ± 0.08
	$t_{I,II}$ = 3.8 df = 102 P <0.000	$t_{I,III}$ = 4.7 df = 150 P <0.000	$t_{I,IV}$ = 6.1 df = 265 P <0.000	$t_{I,V}$ = 7.8 df = 246 P <0.000	$t_{II,III}$ = 0.5 df = 180 P >0.50
	$t_{II,IV}$ = 1.7 df = 295 P >0.70	$t_{II,V}$ = 3.8 df = 276 P <0.000	$t_{III,IV}$ = 1.4 df = 343 P >0.10	$t_{III,V}$ = 4.9 df = 439 P <0.000	$t_{IV,V}$ = 3.1 df = 439 P <0.0.002

As can be seen in these tables, in all cases newborns are characterized by the highest mean number values and a wide range of variability in the distribution of Q-HRs when compared to older individuals. These differences are statistically significant.

The results of a comparative analysis of the frequencies of chromosomal Q-HRs in individuals from different age groups of Kyrgyz and Russian nationality living in Bishkek are presented in tables 2.20.4.3 and 2.20.4.4.

Table 2.20.4.3. Frequency of Q-HRs in seven Q-polymorphic autosomes among Kyrgyz of different age groups (Bishkek).

Location of Q-HRs	Ages				
	Newborns (n = 145)	7-19 years (n = 317)	20-39 years (n = 112)	40-59 years (n = 67)	>60 years (n = 23)
3	145 (0.500)* (31.7)**	251 (0.396) (35.6)	80 (0.357) (29.6)	45 (0.336) (31.0)	16 (0.348) (36.4)
4	16 (0.055) (3.5)	26 (0.041) (3.7)	15 (0.067) (5.6)	6 (0.045) (4.1)	6 (0.130) (13.6)
13	156 (0.538) (34.1)	206 (0.325) (29.2)	90 (0.402) (33.3)	52 (0.388) (35.9)	18 (0.391) (40.9)
14	21 (0.072) (4.6)	30 (0.047) (4.2)	15 (0.067) (5.6)	11 (0.082) (7.6)	0 (0.000) (0.0)
15	55 (0.189) (12.0)	72 (0.114) (10.2)	20 (0.089) (7.4)	14 (0.104) (9.7)	0 (0.000) (0.0)
21	36 (0.124) (7.9)	76 (0.119) (10.8)	33 (0.147) (12.2)	8 (0.059) (5.5)	2 (0.043) (4.5)
22	29 (0.100) (6.3)	45 (0.071) (6.4)	17 (0.076) (6.3)	9 (0.067) (6.2)	2 (0.043) (4.5)
Total Q-HRs	458	706	270	145	44

* Q-HR frequency of the chromosomes analyzed;

** Q-HR frequency as a percentage of the overall number of chromosomal Q-HRs.

Table 2.20.4.4. Frequency of Q-HRs in seven Q-polymorphic autosomes in Russian from different age groups (Bishkek).

Location of Q-HRs	Ages				
	Newborns (n = 37)	7-19 years (n = 67)	20-39 years (n = 115)	40-59 years (n = 230)	>60 years (n = 211)
3	39 (0.527)* (29.3)**	65 (0.485) (38.2)	93 (0.404) (33.1)	186 (0.404) (36.0)	166 (0.393) (41.9)
4	5 (0.068) (3.8)	4 (0.029) (2.4)	3 (0.013) (1.1)	8 (0.017) (1.6)	3 (0.007) (0.8)
13	45 (0.608) (33.8)	57 (0.425) (33.5)	99 (0.430) (35.2)	173 (0.376) (33.5)	127 (0.301) (32.1)
14	9 (0.122) (6.8)	9 (0.067) (5.3)	20 (0.087) (7.1)	31 (0.067) (6.0)	25 (0.059) (6.3)
15	10 (0.135) (7.5)	13 (0.097) (7.6)	24 (0.104) (8.5)	38 (0.083) (7.4)	32 (0.076) (8.1)
21	16 (0.216) (12.0)	15 (0.112) (8.8)	25 (0.109) (8.9)	49 (0.107) (9.5)	23 (0.055) (5.8)
22	9 (0.122) (6.8)	7 (0.052) (4.1)	17 (0.074) (6.0)	31 (0.067) (6.0)	20 (0.047) (5.1)
Total Q-HRs	133	170	281	516	396

* Q-HR frequency of the chromosomes analyzed;
** Q-HR frequency as a percentage of the overall number of chromosomal Q-HRs.

As can be seen in these tables, there is a clear tendency towards a decrease in the frequency of Q-HRs with increasing age in all Q-polymorphic chromosomes, with the exception of autosome 4, which contains the smallest quantity of Q-HRs in human populations. At the same time, the greatest differences between the studied samples are found on chromosomes 3 and 13, containing more than half of all Q-HRs in the genome of human populations. Individuals from all age groups do not significantly differ from each other in terms of Q-HRs in the polymorphic loci of the seven autosomes, corresponding to all our previous results [Ibraimov 1983, 1993; Ibraimov et al. 1996, 1990, 1991, 2000] (see Section 2.20.1).

The results of comparative analysis of the frequency of Q-HRs of different sizes and fluorescence intensities (types of Q-HRs) on autosomes in individuals from different age groups are presented in tables 2.20.4.5 and 2.20.4.6.

Table 2.20.4.5. Fluorescence intensity and size of chromosomal Q-HRs in Kyrgyz of different age groups (Bishkek).

Types of Q-HRs	Ages				
	Newborns (n = 145) I	7-19 years (n = 317) II	20-39 years (n = 112) III	40-59 years (n = 67) IV	>60 years (n = 23) V
QFQ 14	413 (90.2%)	661 (93.6%)	238 (88.1%)	137 (94.5%)	43 (97.7%)
QFQ 15	43 (9.4%)	45 (6.4%)	30 (11.1%)	8 (5.5%)	1 (2.3%)
QFQ 25	2 (0.4%)		2 (0.7%)		
Total Q-HRs	458 (100.0%)	706 (100.0%)	270 (100.0%)	145 (100.0%)	44 (100.0%)
$\chi^2_{I,II} = 4.6$; df = 1; $P < 0.005$; $\chi^2_{I,III} = 0.7$; df = 1; $P < 0.050$; $\chi^2_{I,IV} = 2.5$; df = 1; $P < 0.050$; $\chi^2_{I,V} = 2.7$; df = 1; $P < 0.05$; $\chi^2_{II,I\,II} = 8.1$; df = 1; $P < 0.05$; $\chi^2_{II,IV} = 0.2$; df = 1; $P < 0.05$; $\chi^2_{II,V} = 1.2$; df = 1; $P < 0.05$; $\chi^2_{IV,III} = 4.3$; df = 1; $P > 0.50$; $\chi^2_{III,IV} = 3.6$; df = 1; $P < 0.050$; $\chi^2_{IV,V} = 0.8$ df = 1 $P > 0.50$					

Table 2.20.4.6. Fluorescence intensity and size of chromosomal Q-HRs in Russians of different age groups (Bishkek).

Types of Q-HRs	Ages				
	Newborns (n = 37) I	7-19 years (n = 67) II	20-39 years (n = 115) III	40-59 years (n = 230) IV	>60 years (n = 211) V
QFQ 14	111 (83.5%)	157 (92.4%)	259 (92.21%)	481 (93.2%)	368 (92.9%)
QFQ 15	22 (16.5%)	13 (7.6%)	22 (7.8%)	33 (6.3%)	28 (7.0%)
QFQ 25				2 (0.5%)	
Total Q-HRs	133 (100.0%)	170 (100.0%)	281 (100.0%)	516 (100.0%)	396 (100.0%)

$\chi^2_{I,II} = 5.7$; $\chi^2_{I,III} = 7.2$; $\chi^2_{I,IV} = 12.5$; $\chi^2_{I,V} = 10.4$;
df = 1; df = 1; df = 1; df = 1;
$P < 0.005$; $P < 0.050$; $P < 0.050$; $P < 0.05$; $P < 0.05$;

$\chi^2_{II,IV} = 0.1$; $\chi^2_{III,IV} = 0.3$; $\chi^2_{III,V} = 0.5$; $\chi^2_{IV,V} = 0.1$
df = 1; df = 1; df = 1; df = 1
$P < 0.05$; $P > 0.50$; $P > 0.05$; $P > 0.50$

As can be seen in these tables, in young individuals, and especially in newborns, chromosomal Q-HRs are significantly larger in number and brighter in fluorescence intensity when compared to other age groups.

Tables 2.20.4.7 and 2.20.4.8 show that individuals in younger age groups, especially newborns, in the satellite regions of acrocentric chromosomes, taken separately and in combination, when Q-HRs are present simultaneously on the p13 and p11 loci of these autosomes, the incidence of Q-HRs is higher than in older people.

Table 2.20.4.7. The frequency of Q-HRs localized on satellites (p13) and the short arms (p11) of acrocentric chromosomes in Kyrgyz of different age groups (Bishkek).

Location of Q-HRs	Ages				
	Newborns (n = 145) I	7-19 years (n = 317) II	20-39 years (n = 112) III	40-59 years (n = 67) IV	>60 years (n = 23) V
p13 p11+p11	64 (88.8%) 8 (2.2%)	90 (95.7%) 4 (4.3%)	45 (95.7%) 2 (4.3%)	26 (100.0%0 0	2 (100.0%) 0
Total Q-HRs	72 (100.0%)	94 (100.0%)	47 (100.0%)	47 (100.0%)	26 (100.0%)
$\chi^2_{I,II} = 2.86$; df = 1; P >0.50; $\chi^2_{I,III} = 1.74$; df = 1; P >0.50; $\chi^2_{I,IV} = 3.15$; df = 1; P <0.05; $\chi^2_{I,V} = 0.25$; df = 1; P <0.05; $\chi^2_{II,IV} = 1.14$; df = 1; P <0.05; $\chi^2_{II,V} = 0.09$; df = 1; P <0.05; $\chi^2_{III,IV} = 1.14$; df = 1; P >0.50; $\chi^2_{III,V} = 0.09$ df = 1 P <0.05					

Table 2.20.4.8. The frequency of Q-HRs localized on satellites (p13) and the short arms (p11) of acrocentric chromosomes in Russians of different age groups (Bishkek).

Location of Q-HRs	Ages				
	Newborns (n = 37) I	7-19 years (n = 67) II	20-39 years (n = 115) III	40-59 years (n = 230) IV	>60 years (n = 211) V
p13	24 (85.7%)	18 (85.7%)	53 (94.6%)	55 (93.2%)	38 (95.5%)
p11+p11	4 (14.3%)	3 (14.3%)	3 (5.4%)	4 (6.8%)	2 (5.5%)
Total Q-HRs	28 (100.0%)	21 (100.0%)	56 (100.0%)	59 (100.0%)	40 (100.0%)
	$\chi^2_{I,II} = 1.95$; df = 1; P >0.50;	$\chi^2_{I,III} = 1.28$; df = 1; P <0.05;	$\chi^2_{I,IV} = 1.77$; df = 1; P <0.05;	$\chi^2_{I,V} = 1.69$; df = 1; P >0.50;	
	$\chi^2_{II,IV} = 1.28$; df = 1; P <0.05;	$\chi^2_{II,V} = 1.77$; df = 1; P <0.05;	$\chi^2_{III,V} = i.69$ df = 1 P >0.50		

Tables 2.20.4.9 and 2.20.4.10 show the distribution and mean numbers of Q-HRs in individuals in three age groups of Kazakhs and Russians in Almaty.

Table 2.20.4.9. The distribution and mean numbers of chromosomal Q-HRs per individual on autosomes in Kazakh samples (city of Almaty).

Number of Q-HRs	Newborns (n = 389) I	18-25 years (n = 239) II	>60 years (n = 33) III
0	4		2
1	9	16	5
2	60	36	7
3	85	58	12
4	97	60	6
5	76	44	1
6	38	17	
7	20	7	
8		1	
Total Q-HRs	1520	881	84
Mean number of Q-HRs	3.90 ± 0.07	3.69 ± 0.09	2.55 ± 0.213
Statistics	$t_{I, II}$ = 1.808; df = 626; P = 0.071 $t_{I, III}$ = 5.068; df = 420; P <0.001* $t_{II, III}$ = 4.259; df = 270; P <0.001*		

* These differences are statistically significant.

Table 2.20.4.10. The distribution and mean numbers of chromosomal Q-HRs per individual in autosomes of Russian samples (city of Almaty).

Number of Q-HRs	Newborns (n = 83) I	1-25 years (n = 60) II	>60 years (n = 80) III
0	1		4
1	4	4	6
2	9	7	28
3	14	14	23
4	30	17	14
5	16	13	5
6	6	4	
7	3	1	
Total Q-HRs	321	224	212
Mean number of Q-HRs	3.87 ± 0.156	3.73 ± 0.177	2.65 ± 0.133
Statistics	$t_{I,II}$ = 0.502; df = 141; P = 0.575; $t_{I,III}$ = 5.895; df = 161; P <0.001* $t_{II,III}$ = 4.980; df = 138; P <0.001*		

* These differences are statistically significant.

As can be seen in these tables, in all cases newborns were characterized by the highest value and widest range of variability in the distribution of chromosomal Q-HRs when compared to elderly individuals. These differences are statistically significant. However, as can be seen, there are no statistically significant differences between newborns and students by mean number of Q-HRs in each sample. A more thorough analysis showed that, in fact, there are statistically significant differences between the two age groups in terms of the quantitative content of chromosomal Q-HRs (Table 2.20.4.11). The reason for the lack of difference between these two age groups, as shown in tables 2.20.4.9 and 2.20.4.10, concerns the combination of males and females into a single group. As we showed in the previous section, at the population level, due to the Y chromosome, which has the largest Q-heterochromatin block in the human karyotype, the number of Q-HRs on male autosomes is fewer than on female autosomes [Ibraimov 2014]. Indeed, when we compare newborns and students by sex, it turns out that the differences between these two age groups are statistically significant (Table 2.20.4.11). The same results were obtained for the city of Almaty [Ibraimov et al. 2014b].

Table 2.20.4.11. The distribution and mean number of Q-HRs on autosomes in male and female newborns and 18-25 year old individuals of Kazakh nationality (city of Almaty).

Number of Q-HRs	Newborns Boys (n = 207) I	 Girls (n = 182) II	Students Boys (n = 49) III	 Girls (n = 190) IV
0	3	1		
1	5	4	9	7
2	39	21	12	24
3	47	38	13	45
4	51	46	11	49
5	37	39	4	40
6	18	20	17	
7	7	13	7	
Total Q-HRs	770	750	136	745
Mean number of Q-HRs	3.72 ± 0.102	4.12 ± 0.111	2.78 ± 0.175	3.92 ± 0.104

Statistics
$t_{I,II} = 2.649$; df = 387; P = 0.008*
$t_{II,III} = 5.775$; df = 229; P <0.001*
$t_{III,IV} = 5.119$; df = 237; P = <0.001*
$t_{I,III} = 4.137$; df = 254; P = <0.001*
$t_{II,IV} = 1.313$; df = 370; P = <0.190

* These differences are statistically significant.

The results of a comparative analysis of the frequencies of chromosomal Q-HRs in individuals from different age groups of Kazakh and Russian nationality living in Almaty are presented in Table 2.20.4.12.

Table 2.20.4.12. Q-HR frequencies in seven Q-polymorphic autosomes of Kazakh and Russian samples (Almaty).

Location of Q-HRs	Kazakhs			Russians		
	Newborns (n = 389)	18-25 years (n = 239)	>60 years (n = 33)	Newborns (n = 83)	18-25 years (n = 60)	>60 years (n = 80)
3	369 (47.4)* 24.3**	213 (44.5) 24.2	21 (31.8) 25.0	98 (59.0) 30.5	71 (59.2) 31.9	68 (42.5) 32.1
4	74 (9.5) 4.9	48 (10.0) 5.4	3 (4.5) 3.5	10 (6.0) 3.1	8 (6.7) 3.7	5 (3.1) 2.4
13	480 (61.7) 31.6	268 (56.10) 30.4	26 (39.4) 30.7	104 (62.5) 32.4	72 (60.0) 31.5	73 (45.6) 34.4
14	117 (15.0) 7.7	68 (14.2) 7.7	8 (12.1) 9.3	25 (15.0) 7.8	17 (14.1) 7.4	17 (10.6) 8.0
15	161 (20.7) 10.6	77 (16.1) 8.7	9 (13.6) 10.3	33 (19.9) 10.3	22 (18.3) 10.0	22 (13.7) 10.4
21	196 (25.2) 12.9	108 (22.6) 12.3	10 (15.1) 12.4	30 (18.1) 9.3	16 (13.3) 7.3	18 (11.2) 8.5
22	123 (15.8) 8.0	99 (20.7) 11.2	7 (10.6) 8.8	21 (12.6) 6.5	18 (15.0) 8.2	9 (5.6) 4.2
Total Q-HRs	1520 (195.4) 100.0	881 (184.3) 99.9	84 (126.6) 100.0	321 (193.4) 99.9	224 (186.6) 100.0	212 (132.3) 100.0
Mean number of Q-HRs	3.91	3.69	2.25	3.86	3.16	2.65

* Q-HR frequency for the number of chromosomes analyzed;

** Q-HR frequency as a percentage of the overall number of chromosomal Q-HRs.

As can be seen in this table, with increasing age, there is a tendency towards a decrease in the absolute frequencies of Q-HRs for all Q-polymorphic autosomes. It is noteworthy that this tendency is most pronounced on chromosomes 3 and 13, which contain more than half of all Q-HRs in the human genome at the population level. At the same time, individuals from all age groups do not significantly differ from each other in terms of relative content of Q-HRs, if the Q-HR frequencies are expressed in a percentage of total chromosomal Q-HRs in the population, which fully corresponds to all our previous observations [Ibraimov 1993, 2010; Ibraimov et al. 1986, 1990, 1991, 2000, 2013].

We have found only two papers in the literature that, directly or indirectly, address the variability of chromosomal Q-HRs in individuals of different age groups. Buckton et al. [1976] conducted a study of three populations in Scotland. These consisted of a group of newborns and a cohort of 14-year-old schoolchildren from Edinburgh, while the third sample (65 years and older) was from Barra Island—a relatively isolated location in Scotland. The authors concluded that "rather more variation is found in the Q-band intensity polymophisms: the island population appears to have fewer brilliant and intense variants than do the other two groups, 2.9 per person as compared to 4.2 and 3.9 for the newborn and 14 years old subjects respectively: this may be an age difference rather than a population difference". However, at the same time, they noted that "these figures may reflect a real phenomenon in that the Barra population shows less polymorph variation than the mainland populations; but equally well they may reflect a possible age difference, perhaps in the way that chromosomes from aged individuals respond to technical treatment? These possibilities could be evaluated by studying Barra and mainland population samples with a vertical age distribution". Unfortunately, as far as we know, no such study has yet been carried out in Scotland or elsewhere.

Nazarenko [1987] attempted to identify the "age-related dynamics of the Q polymorphism frequency of chromosomes in an ethnically homogeneous normal human population" in Siberia. The age of the surveyed individuals ranged from 7 to 101. The author divided them into three age groups: 7-19 years; 20-49 years; and 50 years and older. He found the following dynamics of the mean number of Q-variants per individual in the three age groups: 3.67 ± 0.10; 3.45 ± 0.14; and 3.12 ± 0.18, respectively. At the same time, the author offers two explanations for what he discovered. These were: "(1) instability of Q-HRs of chromosomes in ontogenesis, and (2) dependence of adaptability and survival of individuals on the specificity of localization of Q-HRs in chromosomes" [Nazarenko 1987].

Our position on this issue is as follows. Testing the first hypothesis would require the monitoring of a representative sample of individuals from the beginning of their formation as zygotes up to their death. As far as we know, no such study has been conducted. The author's second hypothesis is not supported by our observations—as a rule, an increase in the mean number value of Q-HRs per individual in human populations (*m*) is accompanied by an increase in the absolute frequency of Q-HRs simultaneously for all Q-polymorphic loci, and *vice versa*. In addition, populations with low *m* values have a narrow range of variability in the number of Q-HRs in the genome [Ibraimov 1993, 2000]. The data presented here testify to this (see tables 2.20.4.3, 2.20.4.4, and 2.20.4.12). Exactly the same results can be obtained when analyzing the primary digital data of Buckton et al. [1976] for three age groups.

Our data, together with those of Buckton et al. [1976], indicate that: 1) in human populations there is a clear tendency to a decrease in the amount of chromosomal Q-HRs with increasing age, regardless of racial/ethnic characteristics; 2) of all age groups, the genome of newborns has the most Q-HRs; 3) a decrease in the number of Q-HRs with age does not occur due to the "loss" of Q-heterochromatin on individual loci or chromosomes, but occurs simultaneously in all seven Q-polymorphic autosomes (see tables 2.20.4.3, 2.20.4.4, and 2.20.4.12), which is consistent with our previous observations; 4) a decrease in the number of Q-HRs with age occurs not only in terms of number, but also in size and fluorescence intensity (see tables 2.20.4.5 and 2.20.4.6); 5) the frequency of Q-HRs, localized on the satellite regions of acrocentric chromosomes (see tables 2.20.4.7 and 2.20.4.8) decreases with age, and it is notable that Q-HRs on these loci are more common in individuals who have a greater quantity of Q-HRs in their genome than the population average [Ibraimov and Mirrakhimov 1985].

It should also be noted that a decrease in the number of Q-HRs of chromosomes in human populations among individuals in the older age groups does not seem to occur smoothly. According to our data, by the age of 60, 39.6 % and 47.6 % of chromosomal Q-HRs "disappeared" in the genomes of Kyrgyz and Russians in Bishkek, respectively. Of these, the share of newborn Kyrgyz and Russians was 70.6 % and 58.4 %, respectively. In other words, if the number of chromosomal Q-HRs is indeed small in individuals of older age groups, then it appears that this occurs primarily in the early years of life. It is possible that individuals with a large quantity of Q-HRs in the genome, with all other things being equal, are less likely to live to an advanced age [Ibraimov and Karagulova 2006a].

Thus, the existence of significant differences in the quantitative content of chromosomal Q-HRs in individuals from different age groups can be considered an established fact. However, among the authors of these studies there is no agreement on the nature of such differences, although all of them are supporters of the "selectionist" hypothesis. Some adhere to the point of view that Q-HRs, regardless of their localization on a particular chromosome in the human karyotype, have fundamentally similar structural and functional properties, as follows: "the dose is important for the individual and not the localization of the Q-HRs on a chromosome" [Ibraimov et al. 1986]. By "dose" we mean the total amount of Q-heterochromatic material in the genome of a given individual and not their localization on a particular chromosome in the human karyotype. In other words, this approach is based on the assumption that Q-HRs do not constitute a locus-specific element of the human genome. Those authors who take an alternative approach believe that the deviations of observed Q-HRs frequencies from those theoretically expected (here we are referring to the Hardy-Weinberg rule), which occur in some cases for some Q-polymorphic loci (see Section 2.17) in the population, are not random, but reflect the structural and functional features of Q-HRs localized on different chromosomes and, therefore, they have different selective values [Geraedts and Pearson 1974; Mikelsaar et al. 1975; Nazarenko 1987].

There are no data in the literature indicating the existence of any molecular, structural, or other differences between Q-HRs localized at different loci of Q-polymorphic chromosomes [Prokofyeva-Belgovskaya 1986; Verma 1988; Bhasin 2005]. The results of comparative studies of the patterns of Q-HR distribution in human populations unequivocally show that they behave as a single self-sustaining system, whose behavior can be confidently predicted. For example, the relative share (portion) of Q-HRs at seven Q-polymorphic autosomes at the population level are almost comparable, regardless of racial/ethnic origin or age and sex composition, if the frequencies of Q-HRs are expressed as a percentage of the total number of Q-HRs of chromosomes [Ibraimov 1993, 2011a; Ibraimov et al. 1986, 2000]. We believe that the reduction in the quantity of chromosomal Q-HRs with increasing age is not an ontogenetic process, but most likely the result of natural selection, where individuals whose genome contains more than the mean number of Q-HRs in a population are "screened out" (Table 2.20.4.13).

Table 2.20.4.13. Distribution and mean numbers of chromosomal Q-HRs in Kyrgyz, Kazakh, and Russian samples in various age groups.

Number of Q-HRs	Populations								
	Kyrgyz			Kazakhs			Russians		
	(n = 145) I [1]	(n = 112) II [2]	(n = 23) III [3]	(n = 389) I [1]	(n = 239) II [2]	(n = 33) III [3]	(n = 83) I [1]	(n = 60) II [2]	(n = 80) III [3]
0	4	7	3	4		2	1		4
1	19	20	4	9	16	5	4	4	6
2	23	41	11	60	36	7	9	7	28
3	38	19	2	85	58	12	14	14	23
4	37	16	3	97	60	6	30	17	14
5	16	7		76	44	1	16	13	5
6	5	2		38	17		6	4	
7	3			20	7		3	1	
8					1				
Total	458	270	44	1520	881	84	321	224	212
Mean number of Q-HRs	3.16	2.41	1.91	3.91	3.69	2.55	3.87	3.73	2.65

Statistics	$t_{I,II} = 4.01$; df = 255; P = <0.001* $t_{I,III} = 4.58$; df = 38; P = <0.001* $t_{II,III} = 1.64$; df = 133; P = >0.100	$t_{I,II} = 1.808$; df = 6.26; P = 0.071 $t_{I,III} = 5.068$; df = 420; P = <0.001* $t_{II,III} = 4.259$; df = 270; P = 0.001*	$t_{I,II} = 0.502$; df = 141; P = 0.575 $t_{I,III} = 5.895$; df = 161; P = <0.001* $t_{II,III} = 4.980$; df = 138; P = <0.001*

[1] Newborns; [2]18-25 years; [3]60 years and older.
* These differences are statistically significant.

Therefore, we note again that the cause of the detected difference is the different degree of adaptability of individuals depending on the amount of Q-heterochromatic material in their genome, and not on their localization on one chromosome or another.

2.20.5. Chromosomal Q-HRs in neonates who have died in their early years

As follows from the previous section, the question of whether there are differences in the quantity of chromosomal Q-HRs in the genome of individuals belonging to different age groups in human populations has received an affirmative answer. It turns out that the largest number of Q-HRs are found in the genome of newborns, while individuals from older age groups (60 years and older), regardless of their racial/ethnic origin and the climatic/geographic features of their place of permanent residence, see the smallest number of Q-HRs [Buckton et al. 1976; Ibraimov and Karagulova 2006a; Ibraimov et al. 2014b].

We previously examined the peculiarities of the distribution of Q-HRs in chromosomes of infants who died in their early years. From 2000 to 2003, during a period of extreme economic and social crisis in post-communist Kyrgyzstan, the chromosomal Q-HR content of 145 Kyrgyz and 37 Russian newborns in Bishkek was analyzed [Ibraimov and Karagulova 2006b]. During this time, 22 babies died from various diseases, 17 Kyrgyz and 5 Russian babies, the diagnosis of which was confirmed by autopsy. Table 2.20.5.1 shows the distribution and mean number of Q-HRs on autosomes.

Table 2.20.5.1. The distribution and mean number of Q-HRs per individual in neonates and infants died.

	Kyrghyz		Russians	
Number of Q-HRs	Neonates I	Infants died II	Neonates I	Infants died II
	(n = 145)	(n = 17)	(n = 37)	(n = 5)
0	4			
1	19		3	
2	23		7	
3	38		5	
4	37	9	12	2
5	16	5	7	2
6	5	3	3	1
7	3			
Total number of Q-HRs	458	79	133	24

$\chi^2_{I,II} = 9.03$, df = 2, P<0.05 $\chi^2_{I,III} = 2.18$, df = 2, P>0.05 $\chi^2_{I,IV} = 6.18$, df = 2, P<0.05

$\chi^2_{II,III} = 2.23$, df = 2, P>0.05 $\chi^2_{II,IV} = 0.28$, df = 2, P>0.05 $\chi^2_{III,IV} = 2.26$, df = 2, P>0.05

Mean number of Q-HRs	3.16±0.13	4.65±0.19	3.59±0.23	4.80±0.37

$t_{I,II} = 6.47$, df = 36, P<0.000 $t_{I,III} = 1.52$, df = 180, P> 0.100 $t_{I,IV} = 2.33$, df = 148, P<0.021

$t_{II,III} = 3.55$, df = 52, P>0.001 $t_{II,IV} = 0.37$, df = 20, P> 0.500 $t_{III,IV} = 1.88$, df = 40, P>0.050

As can be seen in this table, newborns are characterized by a very wide range of variability in the distribution of chromosomal Q-HRs. The dead infants, in addition to the high value of the mean number (*m*), are distinguished by an extremely narrow range of Q-HR variability in the sample: the number of Q-HRs in the karyotype varies from 4 to 6, with a value of *m* 4.58 ± 0.23 and 4.8 ± 0.37 in Kyrgyz and Russians, respectively. Table 2.20.5.2 shows the frequency of Q-HRs in seven Q-polymorphic autosomes in the studied samples.

Table 2.20.5.2. Q-HR frequencies in seven Q-polymorphic autosomes in Kyrghyz newborns and infants that died.

Location of Q-HRs	Newborns (n = 145)	Infants that died (n = 17)
3	145 (0.500)* (31.7)**	19 (0.559)* (24.1)**
4	16 (0.055) (3.5)	2 (0.059) (2.5)
13	156 (0.538) (34.1)	27 (0.794) (34.2)
14	21 (0.072) (4.6)	6 (0.176) (7.6)
15	55 (0.189) (12.0)	14 (0.412) (17.7)
21	36 (0.124) (7.9)	6 (0.176) (7.6)
22	29 (0.100) (6.3)	5 (0.147) (6.3)
Total Q-HRs	458	79

* Q-HR frequency of the chromosomes analyzed;
** Q-HR frequency as a percentage of the overall number of chromosomal Q-HRs.

As can be seen in this table, in all the samples studied, Q-HRs of chromosomes occur with a frequency corresponding to those expected on all potentially Q-polymorphic autosomes, i.e., none of the samples found a more preferable localization of Q-HRs on one particular autosome or another, which, in turn, confirms the assumption that the material is not locus-specific in the genome [Ibraimov 1993; Ibraiomov et al. 1986].

It should be particularly noted that, unfortunately, we were unable to trace the fate of all the newborns we studied, since the transitional economy experienced in post-Soviet Kyrgyzstan has forced the majority of the population to constantly change its place of residence in search of work, often going far beyond the borders of the country. Nevertheless, even this small amount of factual material, we believe, is of scientific interest. It would also be of scientific interest to check whether the number of Q-HRs in those children who died in the first days, weeks, months, and years of life was actually greater than in those who survived. If this were the case, then, in the future, it would allow us to establish some reasons for the death of some infants in the first years of life, even in developed countries where modern medical care is available. Beginning in the 1970s, for one reason or another, hundreds and thousands of newborns have been investigated in terms of chromosomal Q-polymorphism [McKenzie and Lubs 1975; Muller et al. 1975; Buckton et al. 1976; Lin et al. 1976], whose subsequent fate could be traced in the present day, at least in economically developed countries.

2.20.6. Chromosomal Q-HRs in patients with alimentary obesity

In recent decades, obesity has become more and more common in economically developed countries with serious medical and social consequences. In addition to the growth in damaging health conditions, like hypertension and cardiovascular disease, overweight people often feel socially inferior. Although significant progress has been made in understanding the functioning of physiological systems regulating bodyweight, we are still far from a complete understanding of the pathogenesis of obesity. This, and the fact that existing methods of treatment and other forms of combating obesity are insufficiently effective, is evidenced by the following fact: more than 90 % of individuals who manage to lose weight by dieting, restore the initial weight soon after finishing their diet [Friedman 2000]. Although much remains unclear, there is growing optimism that aspects predisposing an individual to a positive energy balance will be identified. Most of the current research is focused on the molecular mechanisms regulating appetite, feelings of satiety, energy metabolism, and the process of increasing and differentiating fat cells. It is believed that these processes may be based on genes predisposing an individual to the development of alimentary obesity and a number of geneticists are busy studying DNA sequences for such hypothetical candidate genes [Bray and Bouchard 1997].

Without challenging the importance of these studies, we have chosen a slightly different approach in the search for biological markers that predispose an individual to the development of obesity [Ibraimov 2016b].

This approach is based on the idea of studying the phenomenon of the wide quantitative variability of chromosomal Q-HRs in the genome of human populations in certain purely human pathologies [Ibraimov et al. 2010 a, b]. This quantitative variability of Q-HRs in chromosomes exists only in humans, although the type of constitutive heterochromatin at issue is also present in the genome of two higher primates: *P. troglodytes* and *G. gorilla* [Paris Conference 1971, 1975; Pearson 1973, 1977].

To this end, we studied groups of Kyrgyz and Russian women of reproductive age, suffering from alimentary obesity and without clinically expressed neurohormonal abnormalities. All the surveyed individuals reside in Bishkek, the capital of Kyrgyzstan [Ibraimov 2016b].

The choice of female representatives is primarily due to two reasons: 1) although the variability of the number of Q-HRs in the genome, both in women and men, is determined by the seven Q-polymorphic autosomes, men also have a Y chromosome, which has the largest amount of Q-heterochromatin in the human karyotype and is distinguished by a very high level of polymorphism [Paris Conference 1971, 1975; Ibraimov and Mirrakhimov 1985]; 2) men, at least in Kyrgyzstan, do not tend to monitor their body mass very carefully and as Friedman [2000] rightly noted "... this view is very dependent on the cultural context. In many cultures, obesity is considered to be a sign of affluence and prestige, particularly among those cultures where food is less available", which is generally consistent with the realities of modern day Kyrgyzstan.

The obese women included in our sample suggested that they were desperately fighting every extra gram of fat and would like to have, if not an ideal weight, then at least a weight corresponding to their height. Phenotypically healthy Kyrgyz and Russian women of reproductive age with normal weight were used as controls. Table 2.20.6.1 presents data on the distribution and mean numbers of Q-HRs per individual in the samples examined.

As can be seen in this table, women with obesity, regardless of their ethnic origin, are characterized by a statistically significantly low mean number and a narrow range of variability in the distribution of Q-HRs in the samples when compared to the control. Their similarity in these two main quantitative characteristics of the variability of chromosomal Q-HRs has allowed us to combine them into one group for subsequent comparative analysis. Table 2.20.6.2 shows the values of the absolute and relative frequencies of Q-HRs on seven Q-polymorphic autosomes in the study and control groups.

Table 2.20.6.1. The distribution of the numbers and the mean number of chromosomal Q-HRs per individual in groups of women with obesity and in control samples.

Number of Q-HRs	Obese Females		Controls	
	Kyrgyz (N=56)	Russians (N=44)	Kyrgyz (N=100)	Russians (N=100)
	I	II	III	IV
0	11 (19.6)	5 (11.4)	2 (2.0)	4 (4.0)
1	24 (42.9)	18 (40.9)	11 (11.0)	7 (7.0)
2	19 (33.9)	19 (43.2)	32 (32.0)	24 (24.0)
3	2 (3.6)	2 (4.5)	19 (19.0)	33 (33.0)
4			22 (22.0)	31 (31.0)
5			11 (11.0)	1 (1.0)
6			2 (2.0)	
7			1 (1.0)	
Total number of Q-HRs	68	62	294	283
	χ21,2 = 1. 69 df = 2 $P > 0.50$	χ21,3 = 4.14 df = 2 $P < 0.001$	χ22,3 = 1.78 df = 2 $P < 0.001$	χ23,4 = 14.18 df = 2 $P > 0.95$
Mean number of Q-HRs	1.21 ± 0.11 $t_{1,2}$ = 1.29 df = 99 $P > 0.20$	1.41 ± 0.11 $t_{1,4}$ = 10.41 df = 144 $P < 0.001$	2.94 ± 0.14 $t_{2,3}$ = 8,59 df = 140 $P < 0.001$	2.83 ± 0.11 $t_{3,4}$ = 0.62 df = 189 $P > 0.50$

Table 2.20.6.2. The frequency of Q-HRs in Q-polymorphic autosomes in obese women and the control group.

Location of Q-HRs	Obese females	Control (females)	
	(N = 100)	Kyrgyz (N = 100)	Russian (N = 100)
			I
3	69 (0.345)* (53.1)**	83 (0.415) (28.2)	99 (0.495) (35.0)
4	6 (0.030) (4.62)	17 (0.085) (5.78)	9 (0.045) (3.18)
13	24 (0.120) (18.5)	100 (0.500) (34.0)	99 (0.495) (35.0)
14	7 (0.035) (5.38)	14 (0.070) (4.76)	15 (0.075) (5.30)
15	11 (0.055) (8.46)	27 (0.135) (9.18)	21 (0.105) (7.42)
21	21 (0.105) (7.42)	35 (0.175) (11.9)	17 (0.085) (6.01)
22	7 (0.035) (5.38)	18 (0.090) (6.12)	23 (0.115) (8.13)
Total number of Q-HRs	130	294	283
Mean number of Q-HRs	1.30 ± 0.08	2.94 ± 0.14	2.83 ± 0.11

* Q-HR frequency of the chromosomes analyzed;
** Q-HR frequency as a percentage of the overall number of chromosomal Q-HRs.

A comparative analysis of the studied samples backed up our previous observation that Q-HRs on seven Q-polymorphic autosomes in the genome of human populations are not randomly distributed. The largest number of Q-HRs, as in other previously studied populations, is localized on autosomes 3 and 13 (more than half of all Q-HRs), and the rest are more or less evenly distributed on other Q-polymorphic chromosomes. As can be seen in the table, Q-HRs in obese individuals are found at the expected frequency on all potentially Q-polymorphic autosomes [Ibraimov 1993, 2011a; Ibraimov et al. 1986, 1990, 1991, 1997].

As aptly noted by Friedman [2000]: "because eating is an activity in which we all partake, it is not surprising that almost everyone has an opinion about this subject". Therefore, it is not surprising that basic research on obesity has been very broad [Bray and Bouchard 1997; Schwartz et al. 2000]. However, very little progress has been made in studying the genetic basis of obesity in humans, not to mention the identification of specific structural genes, although several single gene defects causing obesity in animals are known [West et al. 1994; Dragoni et al. 1995; Warden et al. 1995; Chagnon and Bouchard 1996; Gauguier et al. 1996; Galli et al. 1996].

The discovery of leptin has made the greatest impression. In particular, it has been shown that mutations leading to a leptin deficiency cause massive obesity in humans as well as in rodents [Zhang et al. 1994; Montegue 1997]. Leptin can affect energy consumption [Pelleymounter et al. 1995; Friedman and Halaas 1998]. Leptin receptor mutation is also associated with severe obesity [Clement et al. 1998]. These data suggest that obesity is a serious disease and is based on genetic and molecular abnormalities. However, it should be borne in mind that the effect of leptin deficiency is highly dependent on adrenal glycocorticoids. Adrenaloectomized mice of *ob/ob*, *db/db* and other rodent models did not develop obesity. Moreover, this steroid is essential to the development of insulin resistance, in changes in muscle function, and in bone growth [Bray and Bouchard 1997].

Why then do some individuals suffer from obesity while others do not? In response to this question, authors studying the neuroendocrine and molecular aspects of obesity answer that initial sensitivity to leptin may be different in different individuals and that patients with obesity, in general, are leptin-resistant [Pelleymounter et al. 1995; Friedman and Halaas 1998]. The molecular basis of leptin-resistance, in some cases, has been elucidated [Stunkard et al. 1990; Erickson et al. 1996; Fan et al. 1997; Vaisse et al. 1998] and the latest research is connected to the leptin system and the regulation of body weight. It is also recognized that plasticity occurs in this system and factors such as diet, environment, age, and exercise are also important in the pathogenesis of obesity [James 1996]. It has been shown that some environmental factors, such as a diet high in fat, can lead to leptin-resistance, although the mechanism of this phenomenon is not yet clear [Friedman and Halaas 1998]. Our position on this issue will be described in Section 3.3.3.

2.20.7. Chromosomal Q-HRs in alcoholics and drug addicts

Alcoholism and drug addiction are, exclusively, human pathologies. The production and consumption of alcoholic beverages have played a central role in the culture, traditions, and rituals of man. There are significant interindividual differences in the rate of metabolism of ethanol and the same dose can have different effects on different individuals in human populations, even if they are of the same age and sex. It is known that there are significant differences in ethanol metabolism among populations representing different racial/ethnic groups [Erickson 1979]. According to some authors, there is an increasing tendency in the consumption of spirits from low latitudes to high ones [Moor 1964; Keller 1970; Autkiewicz 1987]. A similar pattern has been traced across a single state and region [Miroshnichenko and Pelipas 1991].

Taking into account the known role of alcohol and drugs in energy metabolism, we conducted a study to identify possible links between the tendency of individuals to use both strong alcoholic beverages and drugs and the quantity of chromosomal Q-HRs in their genome, taking into account their racial and ethnic characteristics [Ibraimov 2016c].

Patients undergoing treatment at a drug treatment clinic (Bishkek) were examined for alcohol or drug addiction. It is difficult to call them sick, for, as we have found out, at present, at least in the conditions of the medical service in our country, various interpretations of the concept of "alcoholism" and "drug addiction" do not allow us to adopt a unified system of their classification and diagnosis. Since our sample included individuals who received medical assistance, voluntarily or at the insistence of relatives, we cannot call them sick in a strictly clinical sense. Therefore, we conditionally call them patients with alcoholism and drug addiction.

To study the variability of chromosomal Q-HRs in alcoholics, we selected patients of both sexes belonging to only two ethnic groups (Kyrgyz and Russian). However, the group of drug addicts included all those who received medical care, regardless of their racial or ethnic origin, since drug addiction has appeared in the country as a medical and social phenomenon only in recent (post-Soviet) years. It is apparent that, due to social and psychological factors, women suffering from drug addiction either hide their disease or do not seek medical help. Therefore, only males cosntituted our sample of drug addicts. Table 2.20.7.1 shows the distribution and mean number of Q-HRs of chromosomes in individuals suffering from alcoholism and drug addiction, as well as in control groups (residents of the city of Bishkek).

Table 2.20.7.1. The distribution and mean number of Q-HRs per individual among alcoholics and drug addicts.

Number of Q-HRs	Alcoholics (n = 105)	Drug addicts (n = 100)	Ethnic composition of drug addicts
	I	II	
0	17 (16.2)		1. Kyrgyz 34
1	40 (38.1)		2. Russian 32
2	34 (32.4)	12 (12.0)	3. Uighur 14
3	12 (11.4)	13 (13.0)	4. Korean 18
4	2 (1.9)	36 (36.0)	5.German 2
5		30 (30.0)	
6		9 (9.0)	
Total number of Q-HRs	152	411	
$\chi^2_{I,II} = 114.23$; df = 1; P <0.001			
Mean number of Q-HRs	1.45 ± 0.094	4.1 ± 0.113	
Statistics	$t_{I, II} = 18.17$; df = 203; P <0.000;		

First of all, we note that statistical analysis showed that patients suffering from alcoholism, representing two different ethnic groups, did not significantly differ in the quantitative characteristics concerning the variability of chromosomal Q-HRs, which made it possible to combine them into one group. As is seen in the table, alcoholics are distinguished by the lowest mean number value of Q-HRs per individual in the population (*m*) and the narrowest range of variability in the number of Q-HRs on the chromosomes. Although the highest *m* value was characteristic for people suffering from drug abuse, it should be noted that the range of variability in the numbers of Q-HRs of chromosomes is as narrow as that of alcoholics; however, individuals with large numbers of Q-HRs in the genome prevailed. Table 2.20.7.2 presents data on the comparative analysis of patients with alcoholism with a group of drug addicts in terms of Q-HR frequencies on seven Q-polymorphic autosomes in the genome.

Table 2.20.7.2. Q-HR frequencies in seven Q-polymorphic autosomes in alcoholics, drug addicts, and control.

Location of Q-HRs	Drug addicts (n=100)	Controls		Alcoholics (n = 105)
		Kyrgyz (n = 202)	Russians (n = 556)	
	I	II	III	IY
3	134 (0.670)* (32.6)**	141 (0.349)* (30.7)**	445 (0.400)* (37.3)**	64 (0.305)* (45.1)**
4	10 (0.050)* (2.6)**	27 (0.067)* (5.9)**	14 (0.013)* (1.17)**	2 (0.010)* (1.4)**
13	145 (0.725)* (35.3)**	160 (0.396)* (34.9)**	399 (0.359)* (33.5)**	42 (0.200)* (29.6)**
14	23 (0.005) * (5.6)**	26 (0.064)* (5.7)**	76 (0.068)* (6.4)**	11 (0.052)* (7.7)**
15	34 (0.170)* (8.3)**	34 (0.084)* (7.4)**	94 (0.084)* (7.9)**	5 (0.024)* (3.5)**
21	40 (0.200)* (9.7)**	43 (0.106)* (9.4)**	97 (0.087)* (8.1)**	14 (0.067)* (9.9)**
22	25 (0.125)* (6.1)**	28 (0.069)* (6.1)**	68 (0.061)* (5.7)**	4 (0.019)* (2.8)**
Total number of Q-HRs	411	459	1193	152
Mean number of Q-HRs	4.1 ± 0.113	2.27±0.094	2.15±0.510	1.45 ± 0.094

* Q-HR frequency of the chromosomes analyzed;
** Q-HR frequency as a percentage of the overall number of chromosomal Q-HRs.

As noted above, the sample of drug addicts we studied was not divided according to ethnic origin due to their small number and to ensure adequate statistical analysis. Therefore we limited ourselves to indicating their number. According to our data, the content of chromosomal Q-HRs in the genome of drug addicts, compared to the control, and especially with patients suffering from alcohol abuse, was significantly higher.

As can be seen from this table, there is a well recognized quantitative relationship between Q-HR frequencies on seven autosomes and the mean numbers of Q-HRs in human populations, namely, the trend of Q-HR frequencies on all autosomes depends on the value of *m*, which corresponds to all previous observations [Ibraimov 1993; Ibraimov et al. 1982, 1986, 1990, 1991].

The following aspects are apparent in the data : 1) patients with alcoholism have the lowest quantity of Q-HRs in the genome and do not differ among themselves in all the quantitative characteristics of the variability of chromosomal Q-HRs, despite ethnic origin; 2) drug addicts have significantly more Q-HRs in the genome compared to the control and, in particular, with persons abusing strong alcoholic beverages; 3) however, the examined samples of individuals do not differ in the relative share of Q-HRs in seven Q-polymorphic autosomes, i.e., none of the groups show the preferred localization of Q-HRs on seven potentially Q-polymorphic chromosomes, which, once again, indicates the non-locus specific nature of Q-heterochromatin material in the human genome [Ibraimov et al. 1986, 1990, 1991].

In the literature available to us, it was not possible to find research on biological or genetic markers, confirmed by independent studies, indicating a predisposition to alcoholism or drug addiction in humans. For now, we confine ourselves to citing a few facts that have been repeatedly confirmed, at least in the territory of the former USSR: 1) there is a gradient in the use of strong alcoholic beverages from south to north, across the country, individual republics, territories, and regions [Miroshnichenko and Pelipas 1991]; 2) strong alcoholic drinks are reliably more often consumed by residents of high latitude and high altitude regions, regardless of their national, ethnic, or religious affiliation; and 3) alcoholism in women is more severe than in men. As far as we know, almost the same picture persists in the countries of Europe and North America [Dahlgren 1978; Goodwin 1988]. The possible role of the quantity of chromosomal Q-HRs in the human genome in the pathogenesis of alcoholism and drug addiction is discussed in Part III.

2.20.8. Dose compensation of chromosomal Q-HRs

After long years of observation of the distribution of chromosomal Q-variants in different human populations, both normal and pathological, we have come to the conclusion that the quantity ("dose") of Q-HRs in the human genome is regulated somehow. We conditionally call this phenomenon "Q-heterochromatin dose compensation" and treat it as being analogous to that well-known example from mammalian genetics of the inactivation of one X-chromosome in females. To clarify our observations, we find it necessary to present the data on the phenomenon of X-chromosome inactivation [Ibraimov 2019d].

X-chromosome inactivation ("lyonization") is a process in which one of two X chromosomes in mammalian female cells is inactivated. The DNA of the inactive X chromosome is packaged in transcriptionally inactive heterochromatin. It is believed that X-chromosome inactivation (XCI) evolved to enable dose compensation in marsupial and placental mammals as a way of equalizing X-linked gene expression between XX and XY individuals. XCI mechanisms have been the subject of intensive research for more than half a century and significant progress has been made in this direction (reviewed in Galupa and Heard 2018).

However, several questions remain regarding the causes of XCI. The most important asks what role dose compensation of genes plays. In particular, we believe that XCI does not involve the dose compensation of genes, but rather the dose compensation of the amount of heterochromatin in mammalian genomes. Some further points of interest concerning the causes of XCI in mammals are listed below:

1) Among the higher eukaryotes, including homeothermic animals, XCI occurs only in mammals.
2) Chromosome inactivation does not occur in autosomes even in cases where there is a clear excess dose of genes in the genome. For example, in autosomal trisomy in humans there is no inactivation of the extra chromosome.
3) XCI does not occur in the germ cells of females, where both X-chromosomes are active in all oocytes.
4) It is known that the mammalian X chromosome is very large, harboring >1,000 genes in mice and humans. It is believed that a double dose of some of these genes is clearly problematic because the failure to induce XCI in XX embryos leads to early lethality during development [Takagi et al. 1982; Tada et al. 1993; Buzin et al. 1994; Marahrens et al. 1997]. If all genes are not inactivated on a heterochromatinized X chromosome, then why do mammals need XCI at all?
5) XCI also displays some degree of epigenetic plasticity in pathological contexts such as cancer. For example, the Barr body

appears to be absent in tumors (reviewed in Chaligne and Heard 2014).

Facts supporting our view on the dose compensation of heterochromatin are given in the following. It is known that in somatic nuclei of female mammals, the two X chromosomes display very different chromatin states: one X is typically euchromatic and transcriptionally active, while the other is mostly silent and forms a cytologically detectable heterochromatic structure (Barr body). We believe that the biological meaning of XCI is to compensate for the dose of heterochromatin in the genome of female mammals, in which the total amount of constitutive heterochromatin is usually less than that of males. The following facts speak in favor of such an assumption.

1) The sex chromosomes of mammals are carriers of the largest blocks of chromosomal heterochromatin regions.
2) The amount of heterochromatin in the genome of men has been found to be significantly higher than in women due to the largest block of constitutive heterochromatin on the Y chromosome.
3) There are two groups of data indicating the existence of dose compensation for chromosomal HRs in the human genome. The first group concerns the data on the distribution of the amount of chromosomal Q-HRs on autosomes in women [Ibraimov 2014]. The second group concerns data relating to the relationship between the size of the Y chromosome and the amount of autosomal Q-HRs [Ibraimov et al. 2000]. These data have been obtained for human populations living permanently in different climatogeographic conditions of Eurasia and Africa. Unfortunately, such studies cannot be carried out on chromosomal C-heterochromatin regions (C-HRs) due to the fact that, after C-staining, most chromosomes in the human karyotype cannot be identified.

The first group of data supporting dose compensation of heterochromatin have shown that, at the population level, the total number of chromosomal Q-HRs on autosomes in females is significantly higher than in males (see Section 2.20.3).

The second group of data on the existence of heterochromatin dose compensation is illustrated by the example of the relationship between Y chromosome size and the number of autosomal Q-HRs in human populations of Eurasia and Africa, indicating that there is connection between the mean numbers of autosomal Q-HRs and the sizes of Q-heterochromatin blocks on the Y chromosome [Ibraimov et al. 1986, 2000]. It should be emphasized that, in comparative cytogenetic population studies, only autosomal Q-HRs are considered. Nevertheless, the variability of the largest Q-heterochromatin band in the human genome, localized on the q12 segment of the Y chromosome, has been primarily considered separately from the quantitative variability of autosomal Q-HRs [Geraedts and Pearson 1974; Müller et al. 1975; Buckton et al. 1976; Lubs et al. 1977; Yamada and Hasegawa 1978; Al-Nassar 1981; Ibraimov and Mirrakhimov 1982 a, b, c, 1985; Ibraimov et al. 1982, 1986, 1990, 1991, 2000, 2013].

Table 2.20.8.1 shows the distribution and mean numbers of Q-HRs on autosomes in males having different Y chromosome sizes. Significant decreases in the mean numbers of males with large Y chromosomes were noted. Of interest is the fact that in the group of males with large Y chromosomes a narrowing of the distribution of autosomal Q-HRs was evident, as well as an 'upwards' shift during the greatest range of variability in the number of Q-HRs in the karyotype in groups with medium and small Y chromosomes. Males with large Y chromosomes were characterized by low values in the mean number of Q-HRs per individual and by a low range of variability in the distribution of Q-HRs compared to males with medium and especially small Y chromosomes. These differences are statistically significant.

Table 2.20.8.1. The distribution and mean numbers of autosomal Q-HRs in males with Y chromosomes of various sizes.

Populations	Number of Q-HR	Y ≥ F (n = 30) I	F > Y > G (n = 261) II	Y ≤ G (n = 36) III
Negroids of Africa (Mozambique, Guinea-Bissau, Zimbabwe, Angola)	0			
	1	1		
	2	5	16	
	3	5	42	7
	4	7	64	7
	5	7	66	6
	6	1	46	8
	7	4	20	5
	8		4	2
	9		2	1
	10		1	
Total number of Q-HRs		123	1 220	187
Mean number of Q-HRs		4.10 ± 0.30	4.67 ± 0.09	5.10 ± 0.28
Statistics	$t_{I,II}$ = 1.98; df = 289; P = <0.049 $t_{II,III}$ = 1.61; df = 295; P = <0.108 $t_{I,III}$ = 2.66; df = 64; P = <0.009			
Populations	Number of Q-HRs	Y ≥ F (n = 20) I	F > Y > G (n = 78) II	Y ≤ G (n = 16) III
Steppe Mongoloids (Kazakhs, Chinese)	0	2		
	1	7	11	2
	2	4	26	4
	3	4	21	2
	4	2	11	3
	5	1	4	4
	6		5	1
Total number of Q-HRs		40	220	54
Mean number of Q-HRs		2.00 ± 0.31	2.82 ± 0.15	3.38 ± 0.40
Statistics	$t_{I,II}$ = 2.39; df = 30; P = <0.021 $t_{II,III}$ = 1.32; df = 20; P = <0.194 $t_{I,III}$ = 2.75; df = 32; P = <0.008			

Populations	Number of Q-HRs	$Y \geq F$ (n = 56) I	$F > Y > G$ (n = 280) II	$Y \leq G$ (n = 24) III
Russians	0	10	23	1
	1	21	53	5
	2	14	103	6
	3	7	65	7
	4	4	25	2
	5		9	2
	6		2	1
Total number of Q-HRs		86	611	62
Mean number of Q-HRs		1.53 ± 0.15	2.18 ± 0.07	2.58 ± 0.29
Statistics	$t_{I, II}$ = 3.93; df = 82; P = <0.000 $t_{II, III}$ = 1.34; df = 26; P = <0.183 t I, III = 3.14; df = 37; P = <0.003			
Populations	Number of Q-HRs	$Y \geq F$ (n = 16) I	$F > Y > G$ (n = 125) II	$Y \leq G$ (n = 11) III
Kyrgyz of the Pamirs and Tien-Shan	0	3	11	
	1	5	26	3
	2	6	49	4
	3	2	20	3
	4		14	1
	5		3	
	6		2	
Total number of Q-HRs		23	267	24
Mean number of Q-HRs		1.43 ± 0.24	2.13 ± 0.11	2.18 ± 0.29

Statistics	$t_{I, II} = 2.12$; df = 139; P = <0.036 $t_{II, III} = 0.13$; df = 134; P = <0.900 $t_{I, III} = 1.97$; df = 25; P = <0.060			
Populations	Number of Q-HRs	Y ≥ F (n = 56) I	F > Y > G (n = 215) II	Y ≤ G (n = 26) III
Northern Mongoloids (Chukchi, Yakuts, Khakass, Nenets, Selkups)	0	10	24	
	1	22	50	6
	2	12	71	9
	3	6	46	9
	4	6	17	2
	5		7	
Total number of Q-HRs		88	433	59
Mean number of Q-HRs		1.57 ± 0.16	2.01 ± 0.08	2.27 ± 0.18
Statistics	$t_{I, II} = 2.48$; df = 269; P = <0.017 $t_{II, III} = 1.29$; df = 38; P = <0.198 $t_{I, III} = 2.88$; df = 66; P = <0.005			

It is known that the size of the Q-heterochromatin segment on the long arm of a Y chromosome of even medium size is greater than that found on the Q-HRs on any of the seven Q-polymorphic autosomes in the human karyotype, especially as the morphological variability of the Y chromosome (large, medium, small) is mainly determined by the size of the Q-heterochromatin segment on its long arm. Based on the data presented above, we may assume that the Q-heterochromatin block on the Y chromosome, being the largest Q-heterochromatin segment in the human genome, somehow "restricts" the overall amount of Q-HRs on autosomes in males.

It is likely that, for the same reason, the quantity of autosomal Q-HRs increases in females compared to males within an individual population. In addition, human chromosomal Q-HRs may have a similar role regardless of their location in the karyotype [Ibraimov et al. 2000, 2014]. We believe that the increasing quantity of chromosomal Q-HRs on autosomes in females at the population level is explained by the existence of some evolutionary mechanism that "compensates" the difference in the "dose" of Q-heterochromatin material in the female genome due to the lack of chromosomes in their karyotype, i.e., the Y chromosome, which carries the largest Q-HR segment. As such, there is some mechanism that limits the "dose" of chromosomal Q-HRs in the human genome to a certain level. Indeed, the human karyotype has 25 loci (3 cen, 4 cen, 13 p11, 13 p13, 14 p11, 14 p13, 15 p11, 15 p13, 21 p11, 21 p13, 22 p11, 22, p13, and Yq12), where Q-heterochromatin can potentially be detected. However, as yet no one has been able to find 25 chromosomal Q-HRs in the human karyotype—usually the number varies from 0 to 10 [Yamada and Hasegawa 1978; Al-Nassar et al. 1981; Ibraimov and Mirrakhimov 1985; Ibraimov 2010].

These data give us reason to believe that: 1) Q-heterochromatin on the Y chromosome, constituting the largest block in the human genome, somehow "restricts" the total content of Q-HRs on autosomes in males; and 2) Q-HRs on human chromosomes appear to have a common nature, regardless of their localization in the karyotype.

Lyon [1961] proposed the single active X-chromosome hypothesis to explain the observation that female mice who are heterozygous for X-linked fur color genes display patchy mosaics of two colors. To quote Lyon: "… the heteropicnotic X-chromosome can be either paternal or maternal in origin in different cells of the same animal; … it is genetically inactivated". According to Lyon this mechanism provides dose compensation for X-linked genes as each cell, male or female, has only one X-chromosome that is transcribed.

We believe that Lyon's hypothesis, although flawless in terms of logic, does not fully reflect the essence of XCI. If the only problem was that XCI evolved to enable dose compensation in mammals as a way to equalize X-linked gene expression between XX and XY individuals, then all X-linked genes would be inactivated. The point that we are trying to convey is that: a) X-inactivation is not involved in sex determination, as Lyon stated [1992]; b) the X-chromosome is not inactivated, but rather heterochromatinized in order to compensate for the lack in the female karyotype of the largest block of constitutive HRs found in the male karyotype. The existence of heterochromatin dose compensation, in the example of human chromosomal Q-HRs, we believe to be proven. As such, can we assume that XCI and Q-HR dose compensation in the genome of women are of the same nature? Our answer is yes.

There are a few arguments in favor of such a point of view and some of them deserve closer attention.

1) From a morphological point of view, constitutive and facultative heterochromatin is not significantly different.

2) Once inactivated, the X chromosome is consistently found in close association with the nuclear membrane [Klinger 1958; Hoehn and Martin 1973; Belmont et al. 1986] and/or at the periphery of the nucleolus [Bourgeous et al. 1985; Zhang et al. 2007].
3) The inactivated X chromosome is often found in the nucleolus [Bourgeous et al. 1985], where chromosomal HRs of autosomes 1, 9, and 16 accumulate, as well as the Y chromosome [Schmid et al. 1975], which we consider to be components of cell thermoregulation (CT) involved in the removal of excess heat from the cell nucleus [Ibraimov 2003, 2015a, b, 2017].
4) In embryogenesis, XCI begins at the blastocyst stage, that is, at the stage of multicellularity, when the problem of removing excess heat from the nucleus begins [Ibraimov 2004].

Thus, the cause of XCI is the dose compensation of heterochromatin, rather than of genes, in the genome of female mammals due to the lack of a sex chromosome (the Y chromosome in males) with a large block of constitutive heterochromatin. The heterochromatinization of the euchromatin regions of one of the X chromosomes occurs in order to maintain temperature homeostasis in mammalian female cells [Ibaimov 2019d].

PART III

THE THEORY

In physical sciences, as a rule, theories are based on laws; for example, the laws of motion led to the theory of gravitation. In evolutionary biology, however, theories are largely based on concepts such as competition, female choice, selection, succession and dominance.
—Ernst Mayr (2000. Scientific American, 283(1):69).

3.1. Functions of heterochromatic regions of chromosomes according to the literature

Despite the fact that, by now, quite extensive information has been obtained on the genetic, morphological, and molecular features of heterochromatin in many plant and animal species, as well as in humans, the biological role of chromosomal heterochromatin regions (HRs) remains unclear [Hsu 1975; Prokofieva-Belgovskaya 1986; Zakharov and Tsoneva 1982; Verma and Dosik 1980a; Stahl and Hartung 1981; John 1988; Verma 1988; Bhasin 2005; Ibraimov 2015a]. Attempts to ascertain the functions of these fundamental structures of the genome of higher eukaryotes accompany the entire history of the study of HRs and a large number of plausible hypotheses have been expressed.

As Prokofyeva-Belgovskaya [1986] noted:

> "Most authors who work on this issue consider the arguments only in favor of one definite hypothesis about the role of HRs, whereas at present it has become obvious that the functions performed by HRs are multilateral and manifest themselves, seems to be at all levels of biological organization. In addition, these functions may be related to the organization and functioning of the entire genome, at least, the specific features of the behavior of HRs have not been explained within the concept of the 'autonomy' of the existence of HRs in the genome of higher eukaryotes, which follows from the idea that DNA, part of HRs, is one of the types of so-called 'egoistic' DNA" [Doolittle and Sapienza 1980; Orgel and Crick 1980].

A major feature of HRs, on which hypotheses about their role are based, are that they consist mainly of highly repetitive DNA sequences (this statement is not absolute as there are HRs that do not contain highly repetitive sequences, as well as highly repetitive sequences outside the HRs) [Manuelidis 1982; John 1988]. Thus, the role of HRs is, in principle, irreducible to the role of highly repetitive DNA sequences in the genome. SatDNAs that are part of HRs are not transcribed [Flamm et al. 1969], while at the same time, individual transcribable genes can be part of HRs.

HRs occupy quite definite chromosomal loci that have very important meanings: regions of centromeres and telomeres, as well as regions of the nucleolar organizer regions (NORs) carrying rRNA genes; replication lability, under-replication of chromosomal HRs with polytenisation and the possibility of amplification of the sequences contained in them; and also wide intra-species variability (polymorphism) as well as the evolutionary tightness of HRs in the genome [Prokofyeva-Belgovskaya, 1986].

Regarding attempts to establish the role of chromosomal HRs based on their molecular features, Mikloś [1982] explicitly states that sequence analysis has not brought us any closer to understanding any biological pattern.

A number of authors [Brown 1966; Müller et al. 1975; Jacobs 1977; Herva 1981] suggest that HRs may "not have any function". The basis for such a view concerns the wide variability of HRs on chromosomes, as well as their extraordinary heterogeneity, revealed at the molecular level, the causes of which have not yet been established.

There are a number of hypotheses associated with the possible functioning of HRs in the interphase nucleus. In particular, their possible participation in the formation of the specific structure of the interphase nucleus, which has an important role in maintaining a certain spatial position of the chromosomes relative to each other and the nuclear membrane, has been considered [Comings 1968; Franke and Krien 1972; Parry and Sandler 1974; Schmid et al. 1975; Fussel 1975; Manuelidis 1982; Vogt 1990].

A "bodyguard" hypothesis [Hsu 1975] has also been proposed, relying on the fact that HRs attached to the inner side of the nuclear membrane can serve to protect euchromatic chromosomes functioning in interphase by absorbing chemical mutagens and viruses. This hypothesis suggests that heterochromatin is used by the cell as a protective body to protect euchromatin by forming a layer—a shield—on the outer surface of the nucleolus. Mutagens, clastogens, or even viruses attacking the nucleus make contact with the constitutive heterochromatin, which reflects (absorbs) the attack, thereby protecting the genome in the euchromatic regions of chromosomes. It is possible that the protective force of heterochromatin against penetrating radiation is ineffective, but with respect to chemical agents and viruses, a dense layer of heterochromatin can be a powerful barrier. By "bodyguard", Hsu means not just a means of protection, but a means of protection to the extent that the "bodyguard" serves as a free agent, absorbing the harmful effects entirely or even destroying the threat if necessary. However, the "bodyguard" hypothesis currently lacks sufficient experimental evidence.

A number of authors suggest that the functioning of chromosomal HRs is timed to the process of cell division. Thus, the ability of HRs of chromosomes to participate in non-homologous conjugation can determine the behavior of chromosomes preceding their mating and the formation of a synaptonemal complex [Walker 1971; Yunis and Yasmineh 1971]. This hypothesis is known as the "recognition" hypothesis. However, it brings objections from a theoretical perspective [John and Mikloś 1979], since it remains unclear how non-homologous conjugation can provide "recognition" of homologous chromosomes and, moreover, this property of HRs is more difficult than facilitation of the correct chromosome synapsis in meiosis.

There are two hypotheses on the possible role of HRs in cell metabolism. These are based on the idea that heterochromatin acts as a kind of buffer, regulating the content of overweight nucleotides in cells [Commoner 1964] and histones [Khesin and Leibovich 1976]. From the point of view of these complementary concepts, the very existence of HRs as replicating chromosomal structures is important—they echo Brown's well-known view [1966] on the biological role of heterochromatinthat "it is important not to do nothing".

Some facts indicate the possible transcriptional activity of HRs, like euchromatic regions of chromosomes. One example of a proposed HR action is that the heterochromatic Y chromosome of *Drosophila melanogaster*, at some stages of spermatogenesis, is transformed into a structure resembling a chromosome of the "lamp brush" type [Meyer et al. 1961]. In *Rinchosciaria*, under conditions of a viral infection, puffs are formed at certain sites of the chromosomal HRs, which indicates their transition to an active state [Sanders and Pavan 1972]. In human oocytes, at the diplotena stage, a substance resembling the material of nucleolar organizers is formed, which is constantly associated with the heterochromatin portion of the secondary constriction of chromosome 9 [Stahl et al. 1975].

Gershenson [1933] showed that there is usually no crossing over near chromosome HRs. Based on the comparison of these data with the usual localization of HRs, it was suggested that centromeric HRs, due to the operation of such a mechanism, prevent crossing over in the region of the centromere and thereby fix it in a certain position. The same mechanism can ensure the unity of all blocks of ribosomal genes and prevent crossing over in the sex chromosomes [Yunis and Yasmineh 1971; John and Miklos 1979; Prokofyeva-Belgovskaya 1986], since, as noted above, HRs do not form a synaptonemal complex, which is a necessary precondition for crossing over

Darlington [1937] was the first to attribute heterochromatin an important role in evolution, namely, in speciation through the formation of viable translocations. To date, extensive studies have been conducted to clarify the possible role of chromosomal HRs in the process of speciation in plants and animals. It has been shown that, indeed, the karyotypes of species that do not carry a sufficient amount of HRs are highly stable in phylogenesis [Natarajan and Ahnström 1969]. In a number of rodent species, it has been shown that differences in the quantitative content of heterochromatin may be the only interspecific differences in karyotypes that are otherwise morphologically identical [Pathak et al. 1973; Yosida 1975].

The evolutionary role of HRs can be ensured through various mechanisms. In experiments on *D. melanogaster*, a decrease in the frequency of recombination with a decrease in the content of heterochromatin in the genome was shown [John and Miklos 1979; Prokofyeva-Belgovskaya 1986]. The influence of HRs on the level of population variability and, consequently, on its adaptive capabilities, is considered significant of their important biological role [John and Miklos 1979].

The main result of this work is to highlight that changes in HRs in different species appear to be adaptive in nature, providing them with rapid adaptation to changing environmental conditions [Dyer 1964; Mazrimes and Hatch 1972; Hatch et al. 1976; Borisov 1980].

The idea of a certain selective value of HRs regarding humans has been presented in a number of works. As Soudek [1977] noted, there are at least two assessments of the selective value of HRs: firstly, the presence of HRs is considered to be a selectively neutral or favorable trait and, secondly, there is an increase in risk for carriers of HRs or their offspring. The latter assumption underpins the hypothesis about a possible connection between the variability of chromosomal Q-HRs and a number of pathological conditions of the body. Below, we consider the main works devoted to the search for the negative selective value of HRs, which, according to the authors, can be manifested in individuals through different pathologies.

The high level of interest among researchers is down to the possible connection between the variability of chromosomal Q-HRs in human populations with various deviations in the mental development of individuals. Lubs et al. [1977], studying "white" and "black" children (USA) with different IQ values, found a slightly higher frequency of large C-HRs on chromosome 9 in black children with low IQ; in the Q polymorphism of the chromosomes no differences were found. There were no differences in Q-HRs between mentally retarded and normal individuals according to other authors [Tharapel and Summit 1978; Matsuura et al. 1979]. Differences in the frequency of Q-HRs between healthy individuals and patients with various forms of mental disorder were also not detected [Schwinger and Wehner 1976].

A number of papers have been devoted to the study of the possible relationship of chromosomal HRs and Down's syndrome. Robinson and Newton [1977] studied fluorescent polymorphism in a group of 85 patients with trisomy 21 and found, compared to the control, a significantly higher frequency of brightly fluorescent satellites of chromosome 21. The authors believe that this phenomenon may have a causal relationship to this pathology, increasing the risk of the birth of sick children due to the high probability of the non-disjunction of 21 chromosomes in meiosis I as a result of the mutual attraction of Q-HRs of satellite regions. Mikelsaar et al. [1975] did not notice significant differences between a group of children with Down's syndrome and a group of normal adults in the frequencies of polymorphic chromosome variants. However, in the same year Bott et al. [1975] reported an increase in the frequency of "Q-polymorphism" in patients with Down's syndrome and in their families, compared to the normal population.

Nielsen et al. [1974] indicated an increased frequency of birth of children with various chromosomal abnormalities in carriers of the extreme C-HRs of chromosome 9. There have been other reports on the connection of extreme variants of C-HRs with the birth of children with such anomalies [Halbrecht and Shabtai 1976]. The question of reduced fertility in carriers of heteromorphic variants of chromosomal C-HRs has also been considered [Jacobs et al. 1975].

Different variants of chromosomal HRs have repeatedly been considered in connection to an increase in the frequency of spontaneous abortion of unknown etiology unrelated to chromosomal aberrations in the fetus. "Candidates for the perpetrators" include large Y chromosomes [Patil and Lubs 1977b; Nielsen 1978; Genest 1979] and C-HR inversions of the 9th chromosome [Jacobs et al. 1975]. At the same time, a number of authors have found no such connection [de la Chapelle et al. 1974; Hemming and Burns 1979; Blumberg et al. 1982].

Kruminia et al. [1987] used a different approach to this problem, studying a group of married couples with repeated abortions of unknown etiology in comparison to a group of healthy individuals (both groups were made up of Latvian individuals). The authors attempted to establish a relationship between the frequencies of Q-variants in 12 polymorphic loci of seven autosomes, as well as the mean number of Q-HRs per individual in these two groups. It was concluded that the two samples studied significantly differed statistically in the mean numbers and frequencies of Q-variants in most loci. The authors claim that “the detected reduced content of heterochromatin in phenotypically healthy individuals who do not have repeated spontaneous abortions of unknown etiology in history may indicate the great importance of the organism’s genotype for the manifestation of this negative influence ... some structural variants of chromosomes that are individually harmless to their carriers, combining in the offspring of these individuals, can turn into factors that violate human embryonic development”. From our point of view, the existence of such a relationship is quite possible, but we note that, in this case, it would be more logical to use married couples who do not have a history of repeated spontaneous abortion, rather than 170 healthy individuals as a control.

Kivi and Mikelsaar [1981] used the Q-staining method to determine the possible relationship between the variability of chromosomal Q-HRs and malignant diseases. They examined women with cancer of the ovaries and breasts and found no significant difference between the pathology group and the control in the frequencies of Q-variants, in the mean number of Q-variants per individual, or in the frequency of the pericentric inversion of Q-HRs on chromosome 3.

Of particular interest to many researchers has been the connection between the various clinical manifestations of the variability of Y chromosome length. It has been suggested that the human Y chromosome is genetically inert (in particular, concerning its brightly fluorescent segment), except for two genes: one of them is localized on the short arm, this is the so-called TDF (testis determining factor), and another is found on the non-fluorescent segment of the long arm of the Y chromosome, where the H-Y locus is located, determining the H-Y histocompatibility antigen [Krmpotic et al. 1972; Wachtel et al. 1975; McKusick 1980; McKusick and Ruddle 1977].

Despite the fact that the variability of Y chromosome length is not accompanied by noticeable clinical anomalies, it may be important for the reproductive abilities of the carrier [Patil and Lubs 1977b; Nielsen 1978].

Nielsen and Friedrich [1972] stated that they had found a correlation between the size of the Y chromosome and behavior, in particular, in relation to antisocial behavior. A further report by this group of researchers [Nielsen and Nordland 1975] stated that the existence of such a link could not be confirmed regarding the social adaptation of men, but some correlation was found between long Y chromosomes and the "activity" of men. However, subsequent observations by these authors, as well as other researchers, have shown that there is no correlation between the size of the Y chromosome and behavior, including the "activity" of men [Benezech et al. 1973, 1976; Schwinger and Wild 1974; Brogger et al. 1977].

Yamada et al. [1981] published a very interesting paper on the correlation between the length of the Y chromosome and growth. The sample consisted of 142 Japanese students with an average age of 24. The authors proceeded from the fact that Y chromosome polymorphism is mainly due to the variability of the size of its long arm (Yq). They suggested that the correlation between growth and the length of the Y chromosome should be analyzed separately for the Yq11 and Yq12 segments. According to these authors, the relative length of Yq was 1.63 and the average height of the students was 170.9 cm, fully corresponding to the average height of their peers in Japan. It was found that the growth of the examined men weakly correlated with the total length of the long arm of the Y chromosome, but statistically significantly (at a 5 % confidence level) correlated with the length of the Q-heterochromatic segment (Yq12). No correlation was found between the height and length of the euchromatin segment (Yq11) of the Y chromosome. In other words, tall Japanese students had Y chromosomes with longer Yq12 segments than their shorter peers.

Verma and Dosik [1982] investigated 50 phenotypically normal black Americans aged 25 to 65. According to the authors, this choice was due to the fact that the Y chromosome is longer in black Americans than in Caucasians, and this circumstance should allow for more reliable quantitative results. The average height of the examined black Americans was 172.9 cm and the average length Yq was 1.09. The functional relationship between the two variables, that is, height and Y chromosome length, was calculated by regression. The corresponding statistical analysis showed that there was no significant correlation between height and Y chromosome length in the surveyed black Americans. These data were obtained almost simultaneously, but independently, with Japanese researchers. Therefore, the coincidence of the results of these two observations, which showed the absence of any significant correlation between growth and the total length of the long arm (Yq) of the Y chromosome, is credible, especially since the authors worked with individuals from different races.

Armendares et al. [1983] studied 6 measurements of the body, limbs, and skull, and 6 quantitative characteristics of the heterochromatin segment of the Y chromosome—its absolute and relative length and width, as well as the absolute and relative weight of its magnified image—in 146 healthy Mexican individuals aged 18 to 34. All the studied indicators displayed a normal distribution. The correlation coefficient between them did not differ significantly from 0, except for the absolute width of the heterochromatic segment and the ratio of the length of the upper and lower body ($r = 0.19$; $P < 0.02$).

The above two authors did not find a correlation between growth and the length of the Y chromosome in black Americas, but described a marginal correlation between growth and the length of the Q-HR segment of Yq12 in the Japanese sample. It appears that the variability of Y chromosome length does not really have a significant effect on the growth of men. However, further study is needed to verify these very interesting observations.

A link between the length of the Y chromosome and the non-disjunction of chromosomes in meiosis was also assumed. Large Y chromosomes, as already noted, may be related to spontaneous abortion [Rowley et al. 1963; Kaosaar and Mikelsaar 1973; Patil and Lubs 1977b; Nielsen 1978]; however, this hypothesis was not confirmed in a more thorough study [Schwartz and Cohen 1985]. For example, although an association has been described between the presence of a long Y chromosome in men and the risk of spontaneous abortion in female partners in two families over a number of generations, in another family in the same situation the frequency of spontaneous abortion did not exceed that in the control [Genest and Genest 1985; Westlike et al. 1983; Blumberg et al. 1980]. Verma et al. [1983] indicate that the inconsistency of the data obtained in these studies can be explained both methodologically—the subjectivity of estimating the length of the Y chromosome—and because of the inadequacy of the control.

A number of hypotheses regarding the role of HRs in the regulation of intranuclear and intracellular processes are difficult to reconcile with the wide quantitative variability of these regions as observed in individuals in human populations, especially in relation to Q-HRs. This circumstance, of course, is not evidence of the illegitimacy of currently existing assumptions, but, nevertheless, in itself it requires an explanation in the framework of generally accepted evolutionary ideas.

With regard to attempts to analyze the selective value of human chromosomal Q-HRs in clinical material, data obtained by various groups of researchers on samples of patients with certain groups of pathologies is contradictory and does not allow us to draw any definite conclusions.

Gruzdev [2000] proposed a hypothesis explaining some of the features of heterochromatin, including dense packaging, inactivity in transcription, tendency to aggregation ("stickiness"), and the position effect variation (PEV) in that the DNA molecules in the chromosomal HRs are topologically open and contain single stranded DNA breaks. The formation of heterochromatin is associated with single-stranded breaks in DNA. In euchromatin regions, DNA is topologically closed, supercoiled (torsionally stressed), and hypersensitive to the actions of DNA-relaxing enzymes. According to the author, such cell DNA contains transcribed genes. All this means that different behaviors in the cell cycle and responses to external influences of hetero and euchromatin are associated with the peculiarities of the packing of DNA molecules into them. However, this hypothesis, like the others, does not explain the biological meaning of the existence of wide intra and interpopulation variability in the quantitative content of chromosomal HRs.

Thus, the diversity of roles attributed to heterochromatin highlights our ignorance of its true biological significance, since none of the above hypotheses has garnered sufficient experimental evidence. At the same time, we note once again that all the existing hypotheses relate mainly to the C and not the Q-heterochromatin material of chromosomes.

3.2. Suggestions on the possible biological role of the heterochromatin component of the genome

3.2.1. Methodical and theoretical approaches to the study of chromosomal Q-HR variability

In spite of the fact that research into chromosomal Q-HRs in the genome began almost half a century ago, we still know extremely little of their possible roles in the activity of human life. In the present work, existing methodological approaches in the study of human chromosomal Q-HRs have been analyzed, beginning from empirical observations and going up to analytical approaches, and aiming to detect regularities of Q-HR distribution and possible effects at the population level, both normal and pathological. It appears that this all depends on how we consider the nature of chromosomal Q-HRs, namely, whether they are structurally uniform formations in the genome or their potential effects depend on features of Q-HR localization on this or that chromosome in the human karyotype.

There is a large body of literature devoted to the study of the microscopic structure, molecular composition, methods of identification, and quantitative estimation of chromosomal C-HRs in the human karyotype, as detailed in a number of reviews [Schmid 1967; Prokofyeva-Belgovskaya 1986; Verma and Dosik 1980a; Stahl and Hartung 1981; Ibraimov and Mirrakhimov 1985; John 1988; Verma 1988; Bhasin 2007]. The present work considers these same questions regarding human chromosomal Q-HRs. Therefore, we start with the results of the first empirical observations, which have provided the basis for all subsequent analytical works, and seek out the possible biological role of chromosomal Q-HRs in the human genome.

At the stage of empirical observation, it was shown that, unlike C-heterochromatin, Q-heterochromatin does not always appear. C-heterochromatin is known to be present in all the 46 chromosomes of the human karyotype, varying only in size and location. In contrast, Q-heterochromatin may be completely absent in any of these chromosomes without any appreciable pathological or other phenotypic consequence to the carrier, whereas a complete absence of C-heterochromatin even in one chromosome is an extremely uncommon occurrence (Paris Conference 1971, 1975).

Studies have shown that Q and C-heterochromatin variants sometimes differ, in both location and size, in those chromosomes where they are located in the same regions (being the Y chromosome, chromosome 3, and the short arms of the acrocentric chromosomes) [Verma and Dosik 1980b]. C-heterochromatin is known to account for 15-29 % of the human genome. According to some authors, Q-HR variants of a certain class may be completely absent in a considerable portion of human populations [Yamada and Hasegawa 1978; Al-Nassar et al. 1981; Ibraimov and Mirrakhimov 1982a, b, c, 1985; Ibraimov et al. 1982, 1986]. Thus, there is ample evidence of significant qualitative and quantitative differences between regions of Q and C-heterochromatin.

The chief morphological expression of Q-heterochromatin polymorphism in individuals is the presence, fluorescence intensity, size, and location of Q-heterochromatin in 12 polymorphic loci of seven autosomes (3p11q11, 4p11q11, 13p11, 13p13, 14p11, 14p13, 15p11, 15p13, 21p11, 21p13, 22p11, and 22p13) and in the q12 segment of the Y chromosome. It should be emphasized that there is no individual in a human population who has Q-heterochromatin in all 25 potentially polymorphic loci. The number of Q-heterochromatin variants usually range from 0 to 10 in number

[Ibraimov 2010, 2015a; Ibraimov and Mirrakhimov 1982a, b, c, 1985; Ibraimov et al. 1982, 1986, 2013, 2014].

At the stage of empirical research, the basic questions of identification, calculation, and registration of variants of chromosomal Q-HRs were studied. The discovery of chromosomal polymorphism in humans after Q-staining necessitated the development of a rational and unified system of recording Q-heterochromatin variants, primarily quantitative, for subsequent statistical analysis. Comparison of the frequencies of various types of Q-variants in different populations (normal, pathological, age, ethnicity, etc.) and assessment of their possible adaptive value are the most important problems in any study of human chromosomal polymorphism (see Section 2.11).

Currently, the following quantitative charracteristics of chromosomal Q-heterochromatin polymorphism in human populations are considered relevant: 1) the frequency of Q-HRs in 12 potentially polymorphic loci of seven autosomes. This is usually expressed in the percentage of the number of chromosomes analyzed for each polymorphic locus; 2) the distribution of the number of Q-HRs in a population, i.e., the distribution of individuals with different numbers of Q-HRs in the karyotype regardless of location, which is also reflected in the range of Q-HR variability in the population genome; 3) the mean number of chromosomal Q-HRs per individual, determined by dividing the total number of Q-HRs detected in a given sample by the number of individuals studied; and 4) the size of the Y chromosome, being (a) large (Y = F), (b) medium (F > Y > G), or (c) small (Y = G) [Ibraimov et al. 1990].

At the stage of empirical observation, it was possible to discover: 1) the localization of chromosomal Q-HRs in the human karyotype and methods of their identification; 2) the existence of morphological variants of chromosomal Q-HRs in a human population; 3) the mode of chromosomal Q-HR inheritance across generations; and 4) how to standardize variants of chromosomal Q-HRs (Paris Conference 1971, 1975; ISCN 1978), which is required for comparative population research. These empirical observations made it possible to discover the existence of a wide variability of chromosomal Q-HRs in the genome of human populations. Though these works have not explained the role of chromosomal Q-HRs in the activity of human life, they have allowed us to begin systematic research into the biological role of Q-HRs in the evolution and development of the higher primates.

Currently, the focus is on analytical research. For example, there is still no agreement on the nature of chromosomal Q-HR variability, although all arguments are based on a 'selectionist' hypothesis. One approach, implying that Q-heterochromatin with different locations is basically similar in terms of its structural and functional features, suggests that "of primary importance to an individual is the dose and not the location of Q-variants" [Ibraimov et al. 1986, 1990]. The term "dose" is defined here as the amount of Q-heterochromatin material in the genome regardless of its location in any one chromosome. In other words, this approach is based on an assumption that chromosomal Q-HRs lack locus specificity. Those favoring the alternative approach believe that deviations from expected Q-HR frequencies, observed in any loci, reflect some structural and functional features of these loci and are due either to selection or to non-fortuitous segregation of those chromosomes bearing the given Q-HRs [Geraedts and Pearson 1974; Mikelsaar et al. 1975; Nazarenko 1987].

Let us begin with an analysis of the second approach, which we conditionally name a locus-specific approach. The supporters of such an approach suggest that Q-HRs, localized on different chromosomes in the human karyotype, are different in terms of their structure, properties, and biological effects. This is evidenced by research in which the authors focus on the frequencies of homo (+/+ or -/-) and heterozygotes (-/+) of chromosomal Q-HRs in a population; namely, the agreement of observed homo and heterozygote frequencies in a population with those predicted by the Hardy-Weinberg law. Though there are no direct connections presented in this research, the authors draw an analogy with HbS (sickle-cell trait); nevertheless, judging by the method of the selected statistical analysis it is difficult to exclude such a possibility.

It should be noted that frequencies of homo and heterozygous chromosomal Q-HR variants have not been studied by all investigators in relation to the Hardy-Weinberg law. However, there are many published studies in which the authors have performed a detailed analysis of the agreement of observed frequencies and those predicted by this law of population genetics [Schnedl 1971; Inuma et al. 1973; Mikelsaar et al. 1974; Geraedts and Pearson 1974; Muller et al.1975; Buckton et al. 1976; Van Dyke et al. 1977; Al-Nassar et al. 1981; Ibraimov and Mirrakhimov 1982a, b, c; Ibraimov et al. 1982; Stanyon et al. 1987; Kalz et al. 2005; Decsey et al. 2006]. Agreement has been found in most cases, however, several authors have found some discrepancy between the observed and predicted frequencies of certain Q-HR variants in the seven autosomes.

We have chosen not to examine these observations in detail, since the reasons for these discrepancies between observed homo and heterozygote frequencies and those predicted by the Hardy-Weinberg law are unknown. Therefore, we shall only list some hypothetical explanations for these discrepancies, as suggested by the authors of these studies. These are: 1) the methodological difficulties involved in the calculation of Q-HR variants; 2) too small sample sizes, since the Hardy-Weinberg law is known to be valid only for a large, randomly mated population; and 3) natural selection in the case of excess heterozygotes in a population in certain polymorphic loci [Geraedts and Pearson 1974; Mikelsaar et al. 1974; Müller et al. 1975; Robinson et al. 1976; Tupitsina and Stobetsky 1980].

We found no statistically significant deviations between observed homo and heterozygote frequencies and those predicted by the Hardy-Weinberg law even in those cases where: 1) the sample consisted of no more than 40 subjects; 2) the population had experienced long term exposure to extreme climatic conditions (the Extreme North of Eastern Siberia or the high altitudes of the Pamirs and Tien-Shan); or 3) a relatively high level of inbreeding has occurred [Ibraimov and Mirrakhimov 1982a, b, c, 1985; Ibraimov et al. 1982, 1986]. Jacobs [1977] is of the opinion that it is premature to test the agreement of observed homo and heterozygote frequencies with those predicted by the Hardy-Weinberg law, since methods of calculating the Q-band variants are not sufficiently accurate because of the continuous nature of the size distribution of Q-HRs in a population (for more details, see Section 2.17).

Let us look at the approach in which chromosomal Q-HRs are considered as a single structural and functional system in the human genome and the total quantity, rather than the pattern of localization of Q-heterochromatin segments in the karyotype, is regarded as of paramount importance. If this concerns a specific individual, then the important criterion to estimate the variability of chromosomal Q-HRs is the number (from 0 to 10) of them in the karyotype; if it concerns a population, then the relevant value is the mean number of chromosomal Q-HRs per individual, determined by dividing the total number of Q-HRs detected in a given sample by the number of individuals studied.

The assumption of the absence of Q-HR locus specificity, as revealed by Q-staining, allows us to state on the basis of observation that, with an increase or decrease in the mean numbers of Q-HRs per individual in human populations, there is a simultaneous and proportional increase or decrease in frequency of Q-variants on all Q-polymorphic loci [Ibraimov 1993; Ibraimov et al. 1986]. In particular, it appears that if we arrange all samples according to the mean numbers of Q-HR increase and execute the same operation with the frequencies of Q-variants on seven Q-polymorphic autosomes, then the distribution of populations on the frequencies of Q-variants in these autosomes, as a whole, corresponds to their distribution of the mean numbers of Q-HRs calculated per individual in a population (see Table 2.20.2.11). A similar conclusion was defined by us after the analysis of data from many other researchers [Buckton et al. 1976; Lubs et al. 1977; Kalz et al. 2005], as revealed under the conditions of interpopulation distinctions in mean numbers of Q-HRs in the same laboratory [Ibraimov 1993, 2010].

The following favors the assumption of the absence of locus specificity in chromosomal Q-HR variants: 1) despite the fact that, in the human karyotype, there are 25 loci where chromosomal Q-HRs can potentially be found, in reality the maximum number of Q-HRs does not exceed 10 [Yamada and Hasegawa 1978; Al-Nassar et al. 1981; Ibraimov and Mirrakhimov 1985]; 2) in human populations, the number of Q-HRs in the karyotype usually ranges from 0 to 10 [Ibraimov and Mirrakhimov 1985] without visible phenotypic effects; 3) the distribution of Q-HRs in a population is near normal [Ibraimov et al. 1986, 1990, 1991]; 4) at the population level, the distribution of Q-HRs on seven Q-polymorphic autosomes is uneven, the greatest number of Q-HRs being found on chromosomes 3 and 13 (over 50 %), the rest are more or less evenly distributed on the other five relevant autosomes [Ibraimov 1993, 2010]; 5) human populations do not differ from each other in relative content of Q-HRs on seven autosomes (the portions of Q-HRs on autosomes 3, 4, 13, 14, 15, 21, and 22 are, on average, 25.5 %, 3.5 %, 30.7 %, 8.6 %, 12 %, 10.6 %, and 9.1 %, respectively) [Ibraimov 1993, 2010, 2011a]; 6) the quantitative content of chromosomal Q-HRs in the population genome is best determined by the mean number values of Q-HRs per individual (*m*) [McKenzie and Lubs 1975; Yamada and Hasegawa 1978; Al-Nassar et al. 1981; Ibraimov and Mirrakhimov 1982a, b, c, 1985; Ibraimov et al. 1986, 1990, 1991]; 7) decreases and increases in *m* in a population are due to simultaneous but proportional decreases or increases in the absolute number of Q-HRs on all seven Q-polymorphic autosomes [Ibraimov 1993; Ibraimov et al. 1986, 1990]; 8) there are significant interpopulation differences in the quantitative content of chromosomal Q-HRs in the population genome [Buckton et al. 1976; Lubs et al. 1977; Ibraimov and Mirrakhimov 1982a, b, c; Ibraimov et al. 1982, 1990, 1991; Kalz et al. 2005]; 9) these differences have proven to be related to features of the ecological environment of the place of permanent residence and not to the racial and ethnic composition of the study populations [Ibraimov and Mirrakhimov 1982 a, b, c, 1985; Ibraimov et al. 1982, 1986, 1990, 1991, 2013]; 10) changes in the quantity of Q-HRs in the population genome have a tendency towards a decrease from low to high geographical latitudes, and from low to high altitudes [Ibraimov and Mirrakhimov 1982a, b, c, 1985; Ibraimov et al. 1982, 1990, 1991, 1997, 2013]; 11) both decreases and increases of the *m* value are, as a rule, accompanied by a narrowing or widening of the range of variability in the number of Q-HRs in a population [Ibraimov 1993, 2010]; 12) the segregation of individuals in a human population with different numbers of Q-HRs in the karyotype (from 0 to 10) is due to the fact that Q-HRs are unevenly distributed on seven potentially Q-polymorphic autosomes [Ibraimov 1993, 2010, 2017b]; 13) males in a population differ from each other in the size of the Q-heterochromatin segment of the Y chromosome [Paris Conference 1971, 1975; Yamada and Hasegawa 1978; Ibraimov and Mirrakhimov 1985]; 14) in different age groups, *m* values differ—the greatest number of Q-HRs is characteristic of newborns, while the least number is found in elderly subjects [Buckton et al. 1976; Ibraimov et al. 2014]; 15) in the first days, weeks, months, and years of life, *ceteris paribus*, among healthy children those infants that die often have the greatest number of Q-HRs in the genome [Ibraimov and Karagulova 2006]; 16) individuals that are capable of adapting to the extreme climate of high altitudes (e.g. mountaineers) and the Far North (e.g. oil industry workers on the Yamal Peninsula, Eastern Siberia) have extremely low numbers of Q-HRs in their genome [Ibraimov et al. 1986, 1990, 1991]; 17) individuals with a lower number of Q-HRs in their genome have proven to be prone to alcoholism and obesity, while those with a greater number of Q-HRs are prone to drug addiction [Ibraimov 2010, 2015a]; 18) the Q-HR block on the Y chromosome is the largest in the human karyotype and its size, on average, is twice that of all the Q-HRs on the autosomes taken together [ISCN 1978; Ibraimov and Mirrakhimov 1985]; 19) Q-heterochromatin regions on the Y chromosome, being the largest in the human genome, somehow "restrict" the total content of Q-HRs on autosomes in males. At the population level, the value of *m* is influenced by the quantity of Q-HRs on the Y chromosome—for example, in samples of males with large blocks of Q-heterochromatin on the Y chromosome, the mean number of Q-HRs on their autosomes is lower and *vice versa* [Ibraimov et al. 2000]; 20) there is some mechanism that compensates for the deficiency of Q-heterochromatin material in the female genome due to the lack of a Y chromosome in the female karyotype by increasing the quantity of Q-HRs on autosomes. This pattern persists regardless of age and racial/ethnic characteristics [Ibraimov et al. 2014a]; and finally, 21) individuals in a population truly differ from each other in terms of body-heat conductivity and the level of this difference depends on the quantity of chromosomal Q-HRs in the human genome [Ibraimov et al. 2014b].

Despite the fact that chromosomal C and Q-heterochromatin are defined under the single term "constitutive heterochromatin", they are, undoubtedly, significantly different intrachromosomal structures. There are several significant differences between them: C-heterochromatin is found in the chromosomes of all higher eukaryotes, while Q-heterochromatin is found only in man (*Homo s. sapiens*), the chimpanzee (*Pan troglodytes*), and the gorilla (*Gorilla gorilla*) [Pearson 1973, 1977]. C-HRs are known to be invariably present in all the chromosomes of man, varying mainly in size and location (inversion). Q-HRs can potentially be found only on seven autosomes (3, 4, 13, 14, 15, 21, and 22), as well as on the Y chromosome. Chimpanzees have Q-HRs on five autosomes (14, 15, 17, 22, and 23), while in gorillas they are present on eight autosomes (3, 12, 13, 14, 15, 16, 22, and 23) and on the Y chromosome [Paris Conference 1971, 1975; Chiarelli and Lin 1972; Pearson 1973, 1977; Dutrillaux et al. 1981]. Chromosomal Q-HRs are subject to considerably greater variability in any population compared to C-HRs. Erdtmann [1982] emphasized that "recent analyses ... show a great population and evolutionary stability of C-band homeomorphisms ... From interpopulation comparisons, C-band means show a tendency to maintain a constant amount of constitutive heterochromatin".

The most remarkable aspect in all of this is, in our opinion, the fact that Q-heterochromatin has only been detected in the genomes of the three higher primates. This circumstance previously led some researchers to seek out a relationship between Q-HR variants and human intelligence. Lubs et al. [1977] were unable to find any correlation between IQ and the distribution of Q-HR variants in a group of children aged between 7 and 8. Schwinger and Wehner [1976] were also unable to find any correlation between the frequencies of Q-HRs in seven autosomes of 89 randomly selected normal subjects and 244 patients with various types of mental disorder. Tharapel and Summit [1978] also found no statistically significant differences in Q-HR polymorphism between 200 mentally deficient subjects and 200 controls. The results of Matsuura et al. [1979], obtained in patients with different forms of mental deficiency, did not vary from these observations either.

We have considered chromosomal Q-HRs as a single structural and functional system [Ibraimov et al. 1986]. Usually, a "system" is an aggregate that singles out real or imagined elements in any way from the world [Bertalanffy 1968]. It is possible to consider this aggregate as a biological system, if: 1) each of the elements is indivisible (not in a physical sense, but conditionally); 2) it cooperates with the world as a whole; and 3) the relations existing between elements of the singled out aggregate are set, and in their evolution an unequivocal conformity remains between them so that is there is some orderliness.

The indivisibility of Q-HRs, located in 12 polymorphic loci of seven pairs of autosomes and on the distal part of the Y chromosome, results from the specificity of cytochemical methods used to reveal them [Caspersson et al. 1970; Paris Conference 1971, 1975; ISCN 1978]. The second requirement can basically be reduced to the following: as humans move closer to the northern polar region, as well as with an increase in height above sea level, the amount of Q-heterochromatin decreases in a population's genome, irrespective of the racial/ethnic features of the population, and a reduction of in Q-HRs occurs proportionally on all Q-polymorphic autosomes. Thus, if the elements of the assumed system interact with the surrounding world as a whole unit, the proportional decrease or increase in the quantity of Q-HRs on Q-polymorphic autosomes in a human population gene pool, depending on features of the ecological environment, convincingly testify in favor of the aforementioned representation [Ibraimov 1993, 2010, 2011a, 2015a; Ibraimov et al. 1986].

Concerning the third requirement, it is probably worth noting that: 1) the proportion of Q-HR variants in 12 Q-polymorphic loci of seven pairs of autosomes in human populations, expressed as a percentage of the number of Q-variants revealed in this or that sample, remains comparable, irrespective of the mean number values of Q-HRs calculated per individual in these populations [Ibraimov 1983, 1993, 2010, 2011a, 2017b]; 2) even in those populations where the mean numbers of Q-HRs per individual are very low, Q-heterochromatin does not disappear on autosomes with a low frequency of Q-variants and, moreover, the contribution of Q-high polymorphic and Q-low polymorphic chromosomes in the total pool of Q-heterochromatin material remains quite comparable in different samplings, regardless of racial or ethnic origin or the characteristics of the place of permanent residence [Ibraimov et al. 1986, 1990, 1991, 2013]; 3) the distribution of Q-HRs in a population always comes close to a normal distribution [Ibraimov 1993, 2010]. All this is possible if the studied genetic material in the human genome behaves at the level of human populations as a single self-sustaining structural and functional system [Ibraimov 1993; Ibraimov et al. 1990].

The data, obtained over half a century of empirical observation and analytical study, as described above, allows us to hypothesize that "of primary importance to an individual is the dose and not the location of Q-variants" and to assume that the possible effect of chromosomal Q-HRs on the human body depends on the total amount in the genome [Ibraimov et al. 1986]. As such, we conclude that all of the existing methods for analyzing chromosomal Q-HRs are most fruitful when this type of constitutive heterochromatin is considered as a single self-sustaining structural and functional genetic system in the human genome.

Our ideas about the possible biological role of chromosomal Q-HRs in normal and pathological contexts, as presented further, rely on the methodical and theoretical approaches presented here. First of all, structurally uniform formations in the genome and their possible effects do not depend on Q-HR localization on this or that chromosome in the human karyotype.

However, in terms of everything that is known about the possible role of chromosomal HRs in the human body, we consider **not to be a function, but rather an effect** of the quantity of this genetic material in the genome.

3.2.2. Condensed chromatin and cell thermoregulation

In 1904, Boveri defined chromatin as a substance of the cell nucleus that is transformed into a chromosome during mitosis. Heitz [1928] coined the term heterochromatin to describe and label chromosome segments, and in some cases whole chromosomes, which remain in a condensed state throughout the entire interphase of the cell cycle and therefore appear as a chromocenter in the quiescent nucleus.

Currently, we have extensive information about the features, organization, and properties of chromosomal HRs. The most well-known features of HRs are: 1) they are fixed by evolution in the genome of higher eukaryotes and constitute from 10 to 60 % of the DNA in the genome; 2) they are in a condensed state throughout the cell cycle; 3) they are arranged, as a rule, as short, non-transcribed, tandemly connected, highly repetitive DNA sequences (hrDNA), and, as now has been established, HRs can consist not only of satellite sequences (satDNA), but also of other types of hrDNA; 4) they are localized on the centromeric, telomeric regions of chromosomes, as well as in regions forming nucleolar organizers (NOR); 5) they replicate at the end of the S cell cycle period; and 6) there is a wide inter and intraspecific variability in the quantitative content of chromosomal HRs in human populations.

In talking of the properties of HRs, we usually imply: a) heteropicnosis as a morphological expression of dense packing (condensed chromatin); b) ectopic conjugation of HRs between homologous and non-homologous chromosomes in the interphase nucleus; and c) a high frequency of breaks in areas of HRs or at their border with euchromatin areas.

In recent years, our understanding of the structure and organization of the interphase nucleus has changed drastically. The interphase nucleus has come to be considered primarily as a homogeneous structure that does not have structural specialization, with the exception of the nucleolus. It appears that the interphase nucleus is far from being a homogeneous organelle. It is composed of many substructures and has many functions, each of which is localized in a specific nuclear subdomain. In particular, there is very convincing evidence that each chromosome and its parts are structurally and spatially organized and only occupy discrete domains in the interphase nucleus [Manuelidis 1982; Cremer et al. 1982, 2000; Lichter et al. 1988; Leith et al. 1990; van Driel et al. 1995; Eils et al. 1996].

At the chromosomal level, we can detect the existence of systematic and regular structures and it has been shown that each chromosome mainly occupies its own "territory" in the interphase nucleus [Cremer and Cremer 2001]. Domains with rich and poor gene content exhibit different patterns of orderliness in the nucleus [Croft et al. 1999; Boyle et al. 2001; Cremer and Cremer 2001; Tanabe et al. 2002]. Cremer et al. [2000] believe that, in principle, there are three nuclear complexes: a) an "open" highly orderly complex with chromatin domains containing active genes; b) a "closed" complex with chromatin domains, containing activated genes; and c) a compartment with an interchromatin domain. Such an interchromatin complex can be considered a structural and functional equivalent of the nuclear matrix. Ma et al. [1999] showed that chromosomal territories remain intact even after extraction of the nuclear matrix, but there is some plasticity allowing limited movement and changes in shape, while maintaining a generally high level of organization of chromosomal territories [Dietzel et al. 1995; Eils et al. 1996; Zink and Cremer 1998; Zink et al. 1998]. Ferreira et al. [1997] and Sadoni et al. [1999] distinguish two compartments in the cell nucleus: 1) a late-replicating, transcriptionally inactive compartment consisting of chromatin, forming the G+ and C+ segments on interphase chromosomes and 2) an early-replicating, transcriptionally active complex consisting of chromatin, forming R-segments.

Thus, the accumulated knowledge on the structure of chromatin, excess DNA, and the interphase nucleus inevitably raises at least two questions: 1) why is there so much redundant, "silent" DNA in the chromosomes of higher eukaryotes; and 2) how is the genome organized in the nucleus and how does this genetic organization influence cellular function? These fundamental questions in cell biology are attracting ever increasing attention as the number of higher eukaryotes that have had their genomes sequenced increases.

3.2.3. Arguments in favor of the possible heat conduction role of condensed chromatin in cell thermoregulation

Not yet having direct experimental evidence, I have postulated that the condensed chromatin (CC) of higher eukaryotes may be concerned with intracellular thermoregulation. CC, being the most densely packed material, appears to have the highest heat conductivity in the interphase cell [Ibraimov 2003, 2019e, f, h].

By CC we mean: (1) tissue-specific condensed euchromatin (particular portions of euchromatin condense and differentiate as part of a tissue-specific transcriptional control system); (2) the facultative heterochromatinization of euchromatin (a long-term form of euchromatin repression in which chromosome segments, entire chromosomes, or sometimes even chromosome sets become inactivated and condensed during early development and then remain inactive over many cell generations in all or many somatic tissues); and (3) constitutive HRs [John 1988].

In our opinion, all that is known about chromosomal HRs, interphase nuclei, and redundant DNA does not contradict the idea of a possible heat conduction role of CC in the cytoplasm and cell nucleus. We include the following points:

(1) According to both light and electron microscopy, the nuclear periphery in most cell types is predominantly occupied by heterochromatin, which is closely associated with the lamina and the inner nuclear membrane; and the nucleoli are surrounded by dense chromatin, which also connects the nuclear membrane with one of the nucleoli [Comings 1968; Arrighi and Hsu 1971; Bostock and Sumner 1978; Paddy et al. 1990; Belmont et al. 1993; Marshall et al. 1996; Ferreira et al. 1997; Sadoni et al. 1999]. With the help of the C-technique, CC located on the periphery of interphase nuclei is stained [Arrighi and Hsu 1971; Yunis and Yasmineh 1971].

(2) The membranes of the nuclear envelope serve to compartmentalize the nucleus of higher eukaryotic cells. They are in direct continuity with the endoplasmic reticulum and consist of two concentric bilayers that are joined only at nuclear pore complexes. The outer nuclear membrane shares its proteins and functional properties with the endoplasmic reticulum, whose lumen is continuous with the perinuclear space. The inner nuclear membrane has unique characteristics. It contains a distinct set of membrane proteins [Ellenberg et al. 1997]. Their functions include providing attachment sites for heterochromatin and the nuclear lamina, the

latter of which is a meshwork of intermediate filaments that is associated with interphase chromatin and lines the inner nuclear membrane [Gerace and Burke 1988].

(3) DNA in the eukaryotic nucleus is highly compacted by its association with histones and many other proteins to form chromatin. The formation of the chromosome most probably preceded that of the cell since there are no cells without chromosomes [Lima-de-Faria 1983].

(4) The amount of DNA per cell has been found to bear no relation to the degree of evolutionary complexity (C value paradox). Redundant DNA in most eukaryotic organisms is complexed with proteins in highly compact structures designated as CC. Heterochromatins are a particular case of differential packaging of the chromosome [van Holde 1989].

(5) Heterochromatin has been found to contain more DNA per unit area than euchromatin [Lima-de-Faria 1983].

(6) The centromeres of chromosomes with nucleolus organizing regions (NORs) are consistently associated with nucleoli and in the human genome all rDNA loci are embedded in constitutive heterochromatin. As a result of this linear proximity along the chromosome, nucleoli are always tightly associated with heterochromatin in the interphase nucleus. Genes of 18S, 28S, and 5S RNA, tRNA, and mRNA of histones are located on these sites in different organisms, i.e. all genes, the products of activity of which are necessary in large amounts for general non-specific functions. The transcription of RNA from a DNA matrix occurs in chromosomes during the whole of the interphase and ceases at the mitosis stage.

(7) It has been demonstrated that variable segments (G, Q, and R-bands) are absent in plants and are always present in the chromosomes of higher vertebrates (reptiles, birds, and mammals). In the case of invertebrates, fish, and amphibians, it is difficult to reveal the variable segments. In some insects, some of the segments are equivalent to C-bands, while variable segments appear to be absent [Bostock and Sumner 1978; Prokofyeva-Belgovskaya 1986]. During Q-staining, it can be seen that strongly fluorescent segments are located along the whole length, in all metaphase human chromosomes without exception, but only in individual sites of chromosomes of plants. The banding technique used most successfully in plants corresponds to C-banding and the resulting darkly stained bands probably also represent C-heterochromatin [Vosa 1971]. The difficulties of revealing variable segments in plants, insects, other invertebrates, fish, and amphibians are frequently explained as methodological difficulties. But we believe that they are not connected to the reproducibility of techniques of differential staining and reflect the true state of affairs.

(8) Chromocenters vary by cell type and stage of development, both within and between species [Schmid 1967; Ibraimov 2019e].

(9) There have been observations made of contacts between the nucleolus and HRs of the secondary constrictions of human chromosomes 1, 9, and 16, which do not contain ribosomal cistrons. The frequency of association of acrocentric chromosomes with chromosome 1 grows with an increase in the size of its heterochromatin block [Schmid et al. 1975].

(10) The elucidation of the relationship between temperature and the base composition of DNA of animals can now be better understood in light of the recent discovery that temperature actually changes the genome of higher organisms. Bernardi and Bernardi [1986] made an extensive study of the base composition of the DNA of coldblooded (fish, amphibians, and reptiles) and warmblooded (birds and mammals) vertebrates. Both the noncoding DNA and the sequences that code for proteins (structural genes) turned out to be much richer in guanine-cytosine in warmblooded animals. To test whether this canalization of DNA evolution could be ascribed to the action of temperature, they compared the base composition of related species of fish living in cold (20-25 °C) and warm (37-40 °C) water. Again, guanine-cytosine rich DNA appeared in fish inhabiting warmer water. They concluded that temperature appeared to be a major factor in the canalization of the evolution of DNA. Thus, it is reasonable to expect that in higher vertebrates, such as birds and mammals, thermoregulation ensuring constancy of cell temperature canalizes change in the nucleotide sequences of DNA. As Lima-de-Faria [1983] suggests, the reaction to changes of temperature engages the organism as a whole, but may affect only one or some of its individual components, which have evolutionary trajectories of their own.

(11) Comparison of the contents of satDNA in 12 species and subspecies of the genus *Dipodomys* (kangaroo rats) with their ecological and biological features have shown that the contents of this satDNA are directly connected to the degree of lability of a species and the speed of its adaptation to unexpected changes in environmental conditions; consequently, species rich in satDNA have many subspecies. In contrast, narrow specialized species, which occupy limited niches, have poor quantities of satDNA and few subspecies [Mazrimas and Hatch 1972].

(12) Marked variability of constitutive heterochromatin, not only within but also between species commonly occurs with no phenotypic effect. Even significant changes in the contents of satDNA in a genome have insignificant somatic effects [Miklos 1982].

(13) A new kind of structural heterochromatin—Q-heterochromatin—appeared at a later stage of the evolution of the ancestors of the three

higher primates (*Homo sapiens*, *Pan troglodytes*, and *Gorilla gorilla*) [Paris Conference 1975; Pearson 1973, 1977].

(14) Despite the fact that chromosomal Q-HRs exist in the genome of the three higher primates, their broad quantitative variability is only inherent in human populations [Pearson 1973, 1977; Seuanez et al. 1976; Buckton et al. 1976; Lubs et al. 1977; Al-Nassar et al. 1981; Ibraimov and Mirrakhimov 1982a, b, c, 1985; Ibraimov et al. 1982, 1986, 1990, 1991, 1997; Stanyon et al. 1988].

(15) There is a certain regularity in the distribution of the quantity of chromosomal Q-HRs in the genome of human populations: changes in the amount of Q-HRs in the population genome have a tendency to decrease from low to high geographical latitudes and from low to high altitudes [Ibraimov et al. 1982; 1986; 1990; 1991; 1997] (Fig. 2.13.1).

It is suggested that any serious scientific hypothesis is verifiable. But what conceivable experimental and natural system can be offered to verify the foregoing idea? It might be reassuring if someone managed to show the following: at the change of temperature in the thermostat above or below 37 °C, the speed of transfer of heat from the cytoplasm to the nucleus in a human cell culture depends, for example, on the quantity of chromosomal HRs in the genome of the given individual.

In some sense, nature seems to have done us a favor and conducted part of work for us so that we can understand its "intention". How else would it be possible to explain the following:

(1) In the process of evolution, redundant DNA in the genome of higher eukaryotes appeared as repeated sequences, part of which formed CC in the interphase nucleus. In plants, C-heterochromatin was formed from CC. In the chromosomes of higher vertebrates, in addition to C-heterochromatin, G+ and Q+ bands appeared, which also represent repetitive DNA.

(2) Chromosomal bands, which we identify using differential staining techniques, are best exhibited in homeothermic vertebrates (birds and mammals), rather than in amphibians and reptiles, and worst of all in insects and plants.

(3) Internal conditions (fluctuation of temperature, pH, and osmotic pressure) in animals are more stable than in plants. Amphibians and reptiles have a body temperature partly below the ambient one. Some reptiles, such as the iguana, show a body temperature that is maintained above that of the environment. Birds and mammals control temperature homeostasis within very narrow limits.

(4) Thermoregulation in birds and mammals is mainly achieved by acceleration of the metabolism, stimulated by thyroid hormones, leading to a rise in heat generation. The appearance of the internal environment [Bernard 1872] in evolution, with a high degree of stability, developed mechanisms that allowed these animals to counteract variations in temperature. Birds and mammals developed organ-based control of thermoregulation to regulate their body temperature and have a neural 'thermostat' situated in the hypothalamic region of the brain.

(5) The order Carnivora is quite a unique group among mammals as regards CC features. Identification of chromosomal HRs by means of the C-technique is extremely difficult. In some representatives of the genus *Felix*, the existing techniques of differential staining fail even to identify centromere heterochromatin [Fredga and Mandahl 1972; Pathak and Würster-Hill 1977]. It appears that their known mode of life requires lower heat conductivity inside cells, and possibly inside the whole body.

The possible role of some structural features of chromatin, in particular of the submicroscopic CC segments of the interphase nucleus, in the control of cell thermoregulation has not yet been seriously considered. Though the question of thermoregulation at the level of individual cells has repeatedly arisen, it has not become the subject of systematic and purposeful experimental study [Blatties 1997]. Albeit in a more general or abstract way, just such a question was put forward by Limade-Faria [1983]: "how does the chromosome counteract the environment and death"? He believes that "the chromosome possesses an arsenal of devices that allow it to follow its own rules and to bypass any assault on its integrity, but at the same time it is able to change its structure and function in an ordered way. In this process it is guided only by the physical and chemical principles that were used in its original construction".

In essence, the idea proposed is one reduced to the evolution of the genome structure and the physiology of the whole organism in higher eukaryotes, running in parallel to counteract changes of temperature in the ambient environment for more effective preservation of a constant temperature of the internal environment. The outcomes of such a parallel evolution were: (1) the appearance of different kinds of CC (C and Q-heterochromatin, G+ and Q+ bands, sex chromatin (Barr) body), the effect of which, at the genome level, is generally subject to the laws of physics, and (2) the formation at the organism level of a complex organ-based physiological system of thermoregulation. This is why redundant DNA in the form of HRs has no visible phenotypic expression and bears no specific function because HRs in CC participate in thermoregulation at the level of individual cells; an indirect display of this can be found, e.g., in the wide quantitative variability of chromosomal Q-HRs in human populations permanently living in different climatic and geographical conditions of the earth [Ibraimov 1993, 2010, 2015a, 2017a] (Fig. 2.13.1), as well as in the development of some forms of so-called ‘diseases of civilization’: alimentary obesity, alcoholism, drug addiction, and atherosclerosis [Ibraimov 2016b,c; 2017a,c; 2018a].

Lastly, chromosomes have both internal (repair, recombination, rearrangement, modification, restriction) and external (replication, transcription, packaging, organized movement) molecular activities, which are accompanied, *inter alia*, by some heat output. If, for any reason, the temperature in a nucleus begins to exceed that in the cytoplasm, there is a need for dissipation of surplus heat outside the nucleus. The nucleus has two options to do this: increasing its volume or increasing the conductivity of the nuclear membrane. The first option is limited for obvious reasons. The second option is the more promising. Since the nuclear envelope consists of a double-membraned extension of the rough endoplasmic reticulum, the nuclear membrane cannot essentially change its structure, but it is necessary to remove the surplus heat from the nucleus somehow. Since the proposed idea is based on cell phenomena, in the author’s point of view, nature “found” a very simple and effective solution: it increased its heat conductivity through compression of the internal layer of the nuclear membrane by CC (Fig. 3.2.3.1).

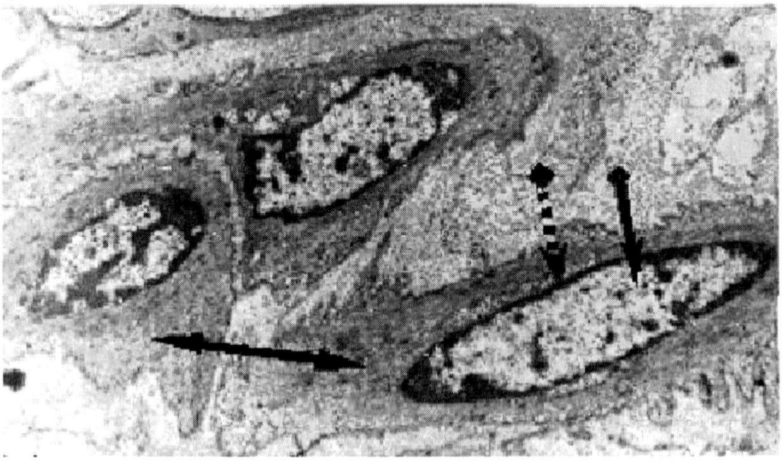

Fig. 3.2.3.1. Electron micrograph of smooth muscle cells of a bull (x 8 000). Condensed (heterochromatin) and decondensed chromatin (euchromatin); ····> peripheric layer of condensed chromatin; → euchromatin; and ↔ cytoplasm can be seen in the nuclei.

Are there any analogues of physical thermoregulation in inanimate systems? In this regard, endothermal organisms (EO) and modern cars equipped with forced cooling system are the most well-known and simplest analogues. Indeed, they have much in common (Table 3.2.3.1). Within the problem discussed, it is important to note that both structures are able to function under the constantly varying temperature conditions of the environment. Thus, the parameters of most elements of both EO and modern cars in physical thermoregulation systems have originally been predetermined by, respectively, evolution and designers. These parameters cannot be changed without detriment to the normal working of the whole system. Nevertheless, at least one element remains in each thermoregulation system to enable adaptation to changes in environmental temperature. These are the cooling system of a car and the amount of chromosomal heterochromatin of EO in the genome, forming a peripheral layer of CC in metabolically active cells, the density of which depends on the quantity and qualitative composition of the constitutive heterochromatin [Ibraimov 2003].

Table 3.2.3.1. Principal analogy of physical thermoregulation systems of endothermic organisms and modern cars.

Elements of the thermoregulation systems	Endothermic organisms	Modern cars
1. The source of heat production.	Metabolically active cells.	Internal combustion engine.
2. The place of heat loss from the source of heat production	Intercellular space.	Engine cylinder cooling jacket.
3. The place of heat transfer.	Peripheral layer of CC around the nucleus.	Engine walls.
4. Cooling system liquid temperature.	Set point maintained by hypothalamus	Set point maintained by thermostat

If this analogy is appropriate, then it is not difficult to imagine EO cells as microengines capable of transferring surplus heat into the intercellular space in a manner similar to the way that internal combustion engines transfer surplus heat into the cooling jackets of cylinders. It is known that the heat in both systems cannot be used as an energy source for normal functioning and therefore this surplus heat energy should be transferred, in due course, out of the cell/engine into circulating liquids.

Finally, we address the qualitative heat loss effect of chromosomal HRs in the nucleus. We wish to show that chromosomal HRs in the nucleus increase the speed of heat loss from the cell nucleus. From a physical point of view, it is clear that the same substance has higher thermal conductivity when it is in a denser state. CC provides the densest areas of the nucleus and consequently their thermal conductivity should be higher than other chromatins.

Let us suppose that all the HRs are localized on the inner surface of the nuclear membrane. In order to create a simple thermophysical model, let us mentally "spread" the HRs along the entire nuclear membrane with an α thick layer. Let us assume that the linear size of the nucleus is equal to r; the HR thermal conductivity rate is equal to λ_1; and the mean thermal conductivity rate of the rest of the nucleus is λ_2. According to our hypothesis $\lambda_2 < \lambda_1$. We also believe that heat production occurs in the entire nucleus.

We assume that the initial ($t = 0$) state is not stationary and then, during $t > 0$, this state relaxes to a stationary one. We feel that the boundary between HRs and the nuclear membrane temperature is set and constant. The mathematical model is described in more detail below.

Let us first present the results of studying this model. With reasonable limitations of the initial distribution of temperature (T) in the cell nucleus, the dependence of temperature upon time with $t > 0$ is described by the expression

$$T = \theta_1(x,t)e^{pt} + \theta_2(x), \tag{1}$$

where

$$p = p_1 = -\frac{\pi^2 \lambda_1}{4\rho_1 c r^2}\left(1 - 2\frac{\lambda_2 a}{\lambda_1 r}\right) \tag{2}$$

if the nucleus contains HRs, and

$$p = p_2 = -\frac{\pi^2 \lambda_1}{4\rho_1 c r^2}\left(1 - 2\frac{a}{r}\right) \tag{3}$$

if there are no HRs in it.

In these formulas: x is the spatial coordinate; ρ1 is the density; and c refers to the specific thermal capacity of the nucleus. It is assumed that $\lambda_1/\rho_1 = \lambda_2/\rho_2$, where ρ_2 ισ is the mean density of the nucleus beyond the area of the Q-HR. Function Θ_2 depends on the spatial coordinate x and describes the distribution of temperature in the nucleus in *t⊠ Ħb*, i.e., after the establishment of the stationary state.

The main result is $|p_1| > |p_2|$, i.e., the presence of Q-HRs increases the speed of heat loss from the nucleus and therefore increases the speed of its cooling and its relaxation towards a stationary state after sudden cooling of the cytoplasm.

Let us describe the mathematical model used. This model contains two one-dimensional, nonstationary, heterogeneous equations of thermal conductivity taking into account heat production in the nucleus. In its physical essence, it describes the compound rod, the left part of which, *a* long, corresponds to the HR, while the right part, *(r-a)* long, models the rest of the nucleus. On the left border of the left rod (i.e., on the boundary between the HR and the nuclear membrane) the temperature is constant. At the contact of two rods (i.e., between the HR and the rest of the nucleus) we have continuous temperature and heat flow density. On the right border of the right rod (i.e., in the center of the nucleus) the density of the heat flow is equal to zero. The initial condition corresponds to a nonstationary state.

The problem of relaxation of the initial state towards a stationary one is solved by using the Laplace transformation on variable *t*, after which the boundary problem for the system of ordinary differential equations is solved. Subsequently, we reverse the Laplace transformation resulting in quadratures such as

$$T(x,t) = \int \tau(x,p) e^{pt} dp$$

The contour of reverse transformation integration is ordinary being the vertical straight line on the complex plane $p = p' + p''$. Function $t(x, p)$ contains the denominator

$$\Delta(p) = \cos(qa)\cos(qr) + \frac{\lambda_2}{\lambda_1}[\sin(qa)\sin(qr)],$$

where $q_2 = \rho_1 cp/\lambda_1 = \rho_2 cp/\lambda_2$. The smallest modulus root $p1$ of the equation $\Delta(p) = 0$ in the assumption that $a << r$ is presented above (2).

The residue in the corresponding function pole $\tau(x, p)$ results in (1). Similarly, we found a solution (3) to $\lambda 1 = \lambda 2$ (in the absence of HRs in the nucleus).

Let us write out the initial equations of the problem, boundary, and initial conditions:

$$\rho_1 c \frac{\partial T_1}{\partial t} = \lambda_1 T_1'' + J_1, \qquad 0 < x < a,$$

$$\rho_2 c \frac{\partial T_2}{\partial t} = \lambda_2 T_2'' + J_2, \qquad a < x < r,$$

Hatch marks indicate the second derivative for x

$$T_1(x,0) = \tau_1(x),\ T_2(x,0) = \tau_2(x),$$

$$T_1(0,t) = T_0 = const,$$

$$T_2'(r,t) = 0,\ T_1(a,t) = T_2(a,t),$$

$$\lambda_1 T_1'(a,t) = \lambda_2 T_2'(a,t),$$

J_1, J_2 = const. heat source densities.

3.2.4. Experimental verification of the cell thermoregulation hypothesis

Certainly, the cell thermoregulation (CT) hypothesis needs to be checked *in vivo* at the cellular level, but we have not yet had such an opportunity. Nevertheless, we have checked this hypothesis at the level of the human organism assuming that CT is the basis for heat conductivity of the cellular part of the body [Ibraimov and Tabaldiev 2007]. However, while determining the quantity of chromosomal Q-HRs in the human genome is a well-established procedure, the same cannot be said about assessing human body heat conductivity (BHC) due to the complete lack of any experience in this regard. In particular, it is still not possible to develop a method to accurately measure the BHC of humans in the way that it is done for homogeneous nonliving objects by physicists.

Through trial and error, we have identified areas of the body (right and left hand and oral cavity) and the thermal load mode (creating an artificial temperature gradient between the left hand and a water bath) that allow us to roughly estimate the level of human BHC (high, medium, and low). Our experience has shown that the most informative factors are (in descending order): a) the time for peak temperature to develop on the surface of the right palm during a thermal load; b) the temperature (*T*) difference between the surface of the right palm and the oral cavity before the thermal load; c) the *T* of the right palm at the moment of peak temperature; and d) the *T* of the right palm at rest. The temperature of the left palm was used only for the preparation of "hot" water, to create a temperature gradient between the arm and the water bath for each person (for more details, see Ibraimov et al. 2014b).

Table 3.2.4.1 shows the relationship between the number of chromosomal Q-HRs in the human genome and the rate of reaction of the body to the controlled thermal load, which was determined by the time (in minutes) to the occurrence of peak temperature on the surface of the right palm.

Table 3.2.4.1. Distribution and mean numbers of chromosomal Q-HRs and time to occurrence of temperature peak on the surface of the right palm.

Number of Q-HRs	1 to 5 minutes (n = 34) I	6 to 10 minutes (n = 75) II	11 minutes + (n = 27) III	Total (n = 136)
2		14	5	19
3	2	12	9	23
4	5	29	8	42
5	14	7	4	25
6	8	3	1	12
7	3	7		10
8	2	3		5
Total	181	306	95	582
Mean number	5.32 ± 0.206	4.08 ± 0.189	3.51 ± 0.209	
Statistics	$t_{I, II}$ = 3.975; df = 107; P = <0.001;* $t_{II, III}$ = 1.656; df = 100; P = 0.101; $t_{I, III}$ = 6.083; df = 59; P = <0.001;*			

* These differences are statistically significant.

As can be seen in this table, there is a statistically significant relationship between the number of chromosomal Q-HRs in the human genome and the reaction of the body to the controlled thermal load. In individuals whose genome contains more chromosomal Q-HRs than the population average, the peak temperature occurs in the first five minutes of the thermal load, and *vice versa*.

The relationship between the quantity of chromosomal Q-HRs and the temperature difference between the surface of the right palm and the oral cavity at rest is shown in Table 3.2.4.2.

Table 3.2.4.2. Distribution and mean numbers of chromosomal Q-HRs and the temperature difference between the surface of the right palm and the oral cavity.

Number of Q-HRs	0.1 °C to 1.0 °C (n = 32) I	1.0 °C to 2.0 °C (n = 71) II	2.1 °C to 3.0 °C (n = 33) III	Total (n = 136)
2		10	9	19
3		12	11	23
4	9	26	7	42
5	9	14	2	25
6	7	4	1	12
7	5	3	2	10
8	2	2	1	5
Total Q-HRs	174	291	117	582
Mean number of Q-HRs	5.44 ± 0.220	4.10 ± 0.168	3.54 ± 0.275	
Statistics	$t_{I, II}$ = 4.607; df = 101; P = <0.001;* $t_{II, III}$ = 1.786; df = 102; P = 0.077; $t_{I, III}$ = 5.349; df = 63; P = <0.001;*			

* These differences are statistically significant.

As can be seen in Table 3.2.4.2, the more chromosomal Q-HRs in the human genome, the smaller the *T* difference between the oral cavity and the surface of the right palm, and *vice versa*. Table 3.2.4.3 shows the relationship between the number of chromosomal Q-HRs in the genome and the value of *T* of the right palm at the moment peak temperature occurs during the controlled thermal load.

Table 3.2.4.3. Distribution and mean numbers of chromosomal Q-HRs and temperature of right palm at the moment peak temperature occurs.

Number of Q-HRs	0.1 °C to 1.0 °C (n = 43) I	1.0 °C to 2.0 °C (n = 65) II	2.1 °C to 3.0 °C (n = 28) III	Total (n = 136)
2		10	9	19
3	4	10	9	23
4	6	31	5	42
5	10	10	5	25
6	12			12
7	6	4		10
8	5			5
Total Q-HRs	240	252	90	582
Mean number of Q-HRs	5.58±0.198	3.87±0.151	3.21±0.208	
Statistics	$t_{I,II}$ = 6.591; df = 106; P = < 0.001* $t_{II,III}$ = 2.474; df = 91; P = 0.015* $t_{I,III}$ = 7.356; df = 69; P = <0.001*			

* These differences are statistically significant.

As shown in the table, there is a statistically significant relationship between the number of chromosomal Q-HRs and the value of *T* of the right palm at the moment of peak temperature occurrence—in individuals with a large number of Q-HRs in the genome, the *T* of the surface of the right palm rises less, and *vice versa.* Table 3.2.4.4 shows a different pattern: the greater the number of chromosomal Q-HRs in the human genome, the higher the *T* of the surface of the right palm at rest, and *vice versa.*

Table 3.2.4.4. Distribution and mean number of chromosomal Q-HRs and the temperature of the surface of the right palm.

Number of Q-HR	Below 35.0 °C (n = 36) I	35.1 °C to 36.0 °C (n = 74) II	36.1 °C and over (n = 26) III	Total (n = 136)
2	5	14		19
3	6	17		23
4	16	23	3	42
5	7	7	11	25
6	2	5	5	12
7		6	4	10
8		2	3	5
Total Q-HRs	139	294	149	582
Mean number of Q-HRs	3.86±0.179	3.97±0.185	5.73±0.239	
Statistics	$t_{I,II}$ = 0.380; df = 108; P = 0.704; $t_{II,III}$ = 5.111; df = 98; P = <0.001* $t_{I,III}$ = 6.395; df = 60; P = <0.001*			

* These differences are statistically significant.

How do we interpret these data? We believe that the time to occurrence of peak temperature on the right palm reflects the rate of heat conductivity, while the value of *T* of the right palm's surface at that moment seems to reflect the quantity of thermal energy in the individual's body. If the peak temperature on the surface of the palm occurs in the first five minutes after the thermal load is applied, then such an individual is considered to have a high BHC, and *vice versa*. In other words, we believe that a person with a high BHC conducts heat through the body quicker and eliminates any excessive through the body shell quicker to better maintain inner body temperature at a constant level.

A statistically significant relationship between the number of chromosomal Q-HRs in the genome and the *T* difference between the oral cavity and the right palm at rest may also characterize the heat conducting ability of the human body—the smaller the *T* difference, the higher the BHC, and *vice versa*. We believe that a smaller *T* difference between the oral cavity and the palm reflects the high heat conductivity of the body, so that such an organism equalizes the *T* difference between the different parts of the body more effectively, thereby successfully avoiding overheating in high temperature conditions. The temperature of the right palm at rest, presumably, also reflects the level of BHC: individuals with a high *T* at the palm may have higher BHC, and *vice versa.*

Heat conductivity caused by the transfer of energy is one of the three phenomena of transfer existing in nature. From the point of view of physics, heat conductivity (HC) involves the transfer of energy from the more heated sites of a body to less heated ones as a result of thermal movement and the interaction of microparticles. HC leads to the equalization of body temperature.

There is nothing new in the idea that the human body possesses some heat conductivity. Nevertheless, it (heat conductivity) has not drawn the attention of physicists or physiologists as an important physical characteristic of the human body. It appears to be connected to the known physical heterogeneity (in a sense, density) of the human body. This is most likely why we did not manage to find any specific method, nor any attempt to estimate the BHC of organisms *in vivo* in the literature [Ibraimov and Tabaldiev 2007; Ibraimov et al. 2010a].

In thermo-physics, the measurement of heat conductivity of solid bodies (e.g., metal) is carried out by determining the heat conductivity coefficient using a calorimetric method. The transfer of heat occurs through a metal rod, the ends of which are placed in a calorimeter with the water and temperatures are taken at $T1$ and $T2$ ($T1 > T2$). It is necessary to estimate HC, where the temperature is lowered to determine the quantity of heat transferred and the time taken to measure the heat conductivity coefficient of the given metal rod. It is obvious that this method of measuring heat conductivity, as applied in thermo-physics, is unacceptable when studying the human body for both technical and ethical reasons. However, we have tried to approximate this problem indirectly, by estimating heat conductivity in a part of the human body. To this end, we had to modify the standard technique to make it acceptable for use with humans [Ibraimov et al. 2014b].

These data show that there are differences in the BHC between individuals in a population. In particular, we were able to show that individuals in a population differ significantly from each other in terms of BHC level. We also found that the level of BHC is affected by sex, age, and the climatic and geographical features of the individual's place of origin. However, the BHC level is not affected by weight, height, arterial pressure, pulse rate, and respiration [Ibraimov and Tabaldiev 2007; Ibraimov et al. 2010a]. In other words, there are some parallels beween the distribution of chromosomal Q-HRs and the variability of BHC at the level of human populations.

Which of the existing biological phenomena underpin this wide human BHC variability in human populations? The first thing that comes to our mind is, of course, the basal metabolic rate, which is a well-known factor in physiology. However, it is also known that the core temperature of those living in the tropics lies within a similar range to those dwelling in the Arctic regions. Apart from that, the basal metabolic rate is influenced by such factors as height, weight, body constitution, pulse rate, and environmental temperature, which contradicts our data [Ibraimov and Tabaldiev 2007; Ibraimov et al. 2010a].

In terms of possible genetic factors, the most appropriate is the amount of chromosomal Q-heterochromatin in human genome. Certainly, the thickness of the peripheral layer of CC around the cell nucleus depends on the total amount of chromosomal C-heterochromatin in the genome. However, we suppose that the packing density (compactization) of CC itself is basically connected with the amount of chromosomal Q-heterochromatin in the nucleus [Ibraimov 2003]. The point is that human populations do not differ significantly in the quantity of C-heterochromatin in their genome [Erdtmann 1982; Ibraimov and Mirrakhimov 1982a]. A wide quantitative variability at the level of human populations is found only in the amount of Q-heterochromatin. The quantitative distribution of chromosomal Q-HRs in a population depends on sex, age, and the peculiarities of the permanent place of residence [McKenzie and Lubs 1975; Erdtmann 1982; Ibraimov and Mirrakhimov 1985; Kalz et al. 2005; Decsey et al. 2006]. It is notable that this regularity has turned out to be very similar to the wide variability of BHC in human populations [Ibraimov and Tabaldiev 2007; Ibraimov et al. 2010a, 2014b]. Thus, it is apparent that human BHC variability depends primarily on the amount of chromosomal Q-heterochromatin in the genome. As the amount of chromosomal heterochromatin does not change in ontogenesis, it is possible that the level of BHC may have a constitutional character, like the color of skin, eye shape, bodily constitution, height, and other innate physical human peculiarities.

3.2.5. Chromosomal Q-HRs in human adaptation, origin, norm, and pathology

The phenomenon of adaptation is at the heart of modern evolutionary biology. Natural selection as a universal mechanism of evolutionary change, first of all, requires the presence of adaptation and a mechanism for its formation. Interest in the phenomenon of biological adaptation arose long before Charles Darwin. His merit lies in the fact that for the first time he developed a natural scientific explanation: adaptation is the key element of natural selection through which poorly and well adapted forms are "screened". It is natural selection that is the primary basis for any biological change and adaptation in a population.

Modern evolutionary theory understands adaptation as a continuous process that is relative, complex, often requiring a compromise, and takes place at different levels of biological organization. Therefore, adaptation is often considered as both a process and a product.

The existence of physiological, morphological, and biochemical changes in the human body as part of the process of adaptation to extreme environmental factors (e.g. cold and hypoxia) has been established. At the same time, the possibility of the existence of some genetic (or more precisely, some genes) changes in the genome of an individual or population has not yet been substantiated by the scientific data. The results of the well-known International Biological Program (IBP) highlight that a number of researchers who have attempted to identify some systematic differences in the distribution of genetically polymorphic systems (blood groups, electrophoretic variants of proteins and enzymes, and anthropogenetic features, etc.) among human populations living permanently in different environmental conditions, including extreme conditions, have not been crowned with success [Baker 1978]. Similar results were obtained in "A Multinational Andean Genetic and Health Program", carried out among indigenous peoples of the highlands and lowlands of South America [Schull and Rothhammer 1977a; Ferrell et al. 1978].

However, the search for specific adaptive genes, in particular to high altitude, continues [Beall 2000; Beall et al. 1994; Ciminelli et al. 2000]. Nevertheless, the fact that there has been successful human adaptation to natural conditions very different to the tropical climate of subequatorial Africa is obvious. The question arises: could a human master the whole oikumene, while remaining a single tropical biological species, without changes in the genome?

From the point of view of genetics, adaptation, like evolution, is a directional and irreversible process. All other (physiological, morphological, biochemical, etc.) processes of adaptation are considered to be reversible, non-inherited adaptive changes in the body and are considered to relate to either acclimation or acclimatization. Adaptation, as well as evolution, involves the participation of three fundamental biological processes: variability, heredity, and natural selection. Of these, hereditary variability is the most important. Without it, there is no selection and, therefore, there is no adaptive evolution.

H. sapiens, as a species, was formed in subequatorial Africa and, remaining a single tropical species, it managed to master the whole oikumene, continuing to expand its zone of permanent residence and abundance. The question of how exactly humans managed to conquer all the lands of the Earth remains generally unclear. *H. sapiens* has always had to adapt to the climate and its changes, starting from the time of its appearance in conditions that were quite complex and which ultimately brought to life such perfect physiological plasticity. The question remains as to what underlies such plasticity.

It is believed that beings indistinguishable from modern man appeared in Africa about 100,000 years ago, 30-40,000 years ago they reached Europe, Asia and Australia, and more than 20,000 years ago they reached North and South America. In speaking of modern man—*Homo sapiens sapiens*—we usually have in mind three major features that distinguish him from other hominids: high physiological plasticity, the structural features of the upper limbs, and the ability to engage in conceptual thinking. Modern man also has such features as bipedal walking, a large neocortex, reduced anterior teeth dominated by molars, unique sexual and reproductive behavior, and the development of material culture, etc. [Harrison et al. 1977]. It is clear that this is an unordered and far from complete list characteristic of *H. s. sapiens* and can easily be supplemented.

It is obvious that the modern man, already formed, adapted to high altitudes and latitudes. How our distant ancestors adapted to the changing climatic conditions of the Middle and Late Miocene in Africa, we may never be able to find out. But we believe that the adaptation strategy of *H. sapiens* can still be understood in terms of the changes that have occurred in the genome of modern man in the process of his mastering the extreme natural conditions of high altitudes and the Far North. This process, which lasted for several millennia, did not lead to serious microevolutionary processes, so that one may speak about the appearance, if not of a new species, then at least of a subspecies. This indirectly indicates that the strategy of human genetic adaptation is likely to be different from that of other animals.

We hold some reservations here in that we are only talking about that part of the human genome—euchromatin or heterochromatin, or more precisely, the gene or non-gene—that is involved in the process of genetic adaptation. The need for such a reservation is justified by the well-known fact that, as a rule, genetic adaptation refers to structural genes.

For various reasons, almost all researchers fall on the side of the gene component of the genome being the most important driver of human adaptation. As such, we summarize here our "anti [gene]" thesis:

1) The structural genes occupy only a small fraction of the DNA in human chromosomes. Only 3-5 % of DNA in the human genome is sufficient for at least 100,000 specific structural genes, and these are enough to ensure the functioning of such a complex organism as *H. s. sapiens*.
2) In humans, not a single protein or enzyme has been found that, in one form or another, is completely absent in other animals.
3) *H. sapiens* was, and remains, a single tropical biological species.
4) For more than a million years of their evolution, our ancestors existed in lowland tropical Africa and never lived in conditions of high altitude or the climate of the Far North.
5) Fully formed man originated in Africa about 50,000 years ago, and permanent populations living at high altitude and the Far North have existed for an even smaller period of time, no more than a few thousand years.
6) Almost all high altitude populations, with the exception of those residing in the Ethiopian Highlands, are representatives of the Mongoloid or Proto-Mongoloid (North and South America) races. This suggests that the age of even the most ancient, permanent high altitude populations cannot exceed 20,000 years, bearing in mind that modern human races had not yet formed at this time.
7) Recent studies show that genetic differences between the currently existing human races are almost negligible. For example, a comparison of the length of human and chimpanzee chromosomes has shown that the total amount of chromosomal material in these two species is very close: they differ only in the distribution of heterochromatic regions, and in a number of pericentric inversions and translocations [Paris Conference 1971, Suppl. 1975; Warburton et al. 1973; Yunis and Prakash 1982]. Amino acid composition and amino acid sequences in 44 proteins of humans and chimpanzees differ by no more than 1 % [King and Wilson 1975]. DNA hybridization experiments have shown that at least 98 % of non-repetitive human DNA in chimpanzees is identical—a value that fully admits the possibility of a viable hybrid. Such a significant degree of similarity makes it difficult to reasonably explain the large biological differences observed between these closely related biological species.

8) Since man, in his evolutionary past, never lived at high altitude, but, nevertheless, has managed to master all the high altitude provinces of the Earth over the last few thousand years, the inevitable question arises, when could hypothetical high altitude adaptive genes have appeared? The problem is further complicated by the fact we do not know whether such genes would arise (if they exist at all in nature)—on a single occasion even before humans began to penetrate the high altitude zones (as a preadaptation), or appearing each time human populations tried to adapt to the extreme natural factors of life at high altitudes (i.e., to hypobaric hypoxia)? If we recognize the possibility of the reemergence of specific adaptive genes in the human genome, do such reemergences relate to the same genes, or did such changes occur in different parts of the genome? Leaving open the question of the possibility of the existence of such a highly mutable gene or genes, let us ask another question: how did such favorable mutations manage to spread in the genome of all the inhabitants of high altitude populations? After all, these populations, even within the same mountainous province, are often isolated by vast geographical distances in the form of formidable high mountain passes and other obstacles. The problem is further complicated by the fact that the aborigines of the high altitude provinces of Asia and South America never "exchanged genes" in historical times, and their ancestors were, respectively, Mongoloid and Proto-Mongoloid tribes that mastered these mountain areas on different continents and geographic latitudes at different times, having different ethnic compositions and ecological experiences in the past.
9) The existence of a certain "high altitude ceiling" (about 4,200 m above sea level) for humanity's permanent residence.
10) Extremely high infant mortality in high altitude populations indirectly indicates the absence of effectively functioning specific adaptive structural genes in the genomes of highlanders. Indeed, with a modern home, clothing, food, and technical means, a human can exist for a long time in the Far North and even in conditions of weightlessness. This is a purely technological and economic problem. However, even modern scientific and technological advances cannot protect humans from hypobaric hypoxia at high altitudes.
11) If we believe that the human settlement of Earth was encouraged primarily by three qualitative features—high physiological plasticity, morphological features of the upper extremities, and the

ability for conceptual thinking—then when did man become the owner of such powerful aromorphoses? What were the prerequisites for their formation? Did they arise before or after humans began to explore new natural environments outside the African savannahs? Can we assume that these three human features were produced by specific structural genes? If so, when and why did they arise in humans?

12) In the human genome, there are a lot of structural genes and complex gene complexes with high rates of polymorphism in human populations. But this is not unique, being a characteristic shared by the entire animal kingdom. The problem lies elsewhere: the functions of these genes are too simple and rather ancient, and they are probably not directly related to the complex physiological and behavioral reactions of the human body to cold or hypobaric hypoxia.
13) Since man in his evolutionary past, at least before the formation of his modern appearance, did not encounter cold and hypoxia to such an extent that they exerted selective pressure, it is difficult to imagine the emergence in his genome of some specific structural genes or gene complexes of adaptation to cold and hypobaric hypoxia.
14) For an adequate response to the selective pressure of cold and hypobaric hypoxia (not to mention other known harmful natural physical factors of high altitudes and the Far North), such a complex organism as man could hardly restrict himself to the responses of one or several structural genes known for their conservatism.
15) Finally, to the question "what do genes actually do?" a leading evolutionist, Lima-de-Faria [1988], responds: "Genes do not change constantly, as suggested earlier, many genes remained unchanged from bacteria to humans. There are many such genes and 28S and 18S ribosomal RNA can be cited as an example. … The gene creates neither form nor function. Gene products only channel the response in one direction or another. ... Thanks to the gene, the cell has acquired some new features: the ability to repeat, additional ordering, acceleration of events, the consolidation of alternatives and increased combinational ability". Speaking of the role of genes in human evolution, he writes: "Human evolution seemed to depend mainly on changes at the level of regulatory DNA, and not at the level of structural genes. The role of structural genes in evolution seems more modest".

It seems to us highly probable that there was another strategy at play in humanity's genetic adaptation to cold and hypobaric hypoxia. Humans used the non-genic part of their genome, in the form of noncoding DNAs, which are known for high mobility and non conservativeness in individual development and evolution. Our long term studies aimed at discovering the genetic basis of the adaptation of human populations to the extreme climatic conditions of high altitude in Central Asia and the Extreme North of Siberia show that chromosomal Q-HRs in condensed chromatin are among the sought-for genetic materials that ensured the successful settlement by man of all the lands of the Earth [Ibraimov 1993, 2010, 2011, 2015].

Based on the above, the adaptation of humans to extreme environmental conditions can be presented in the following way. Human adaptation to different climatogeographic conditions has two fundamental features: a) only man managed to master all the lands of the Earth while remaining a single, tropical species (*H. s. sapiens*); b) unlike animals, the human settlement of all climatogeographic provinces, including extreme zones (the Far North and high altitudes), occurred over a very short period of time. While it has taken hundreds of thousands to millions of years for animals to adapt to their permanent habitat, it only took humans from 700 to 1,000 years (in the Pamirs and Tien-Shan) to 10,000-20,000 years (in the Andes and Tibet).

To date, numerous studies have been carried out on the genetic basis of human adaptation to high altitude climates. First, we briefly present the results of those studies aimed at detecting the gene bases of human adaptation. The history of the study of genetics in human adaptation to extreme natural conditions, including high altitudes, begins with the International Biological Program (IBP). However, the results of the IBP indicate that the attempts of a number of investigators to reveal systematic differences in the distribution of genetically polymorphic systems (in terms of blood group, electrophoretic variants of proteins and enzymes, anthropological traits, etc.) between human populations living permanently under different environmental conditions of the terrestrial globe were unsuccessful [Baker, 1978]. Almost the same results were obtained during the realization of "The Multinational Andean Genetic and Health Program" studying indigenous peoples of the high and low-altitude regions of South America [Schull and Rothhammer 1977a, b].

However, the search for specific genes of adaptation, particularly to high altitude climates, continues, now using new methods of DNA analysis. Researchers have applied different genomic strategies to samples to uncover these mechanisms. Most studies seeking to understand the genetic basis of adaptation to high altitude rely on statistical associations between genes and phenotypes. While this is an important first step towards understanding the complexity of phenotypes, functional tests are needed to uncover causal relationships.

Here, as an illustration, we present the results of only a few pieces of research, while limiting ourselves to referencing review articles. So, for example, Hancock et al. [2011] scanned the human genome using data for about 650,000 variants in 61 worldwide populations to look for correlations between allele frequencies and 9 climate variables and found evidence for adaptations to climate at the genome level. In addition, the authors detected compelling signals for individual SNPs involved in pigmentation and immune responses, as well as for pathways related to UV radiation, infection and immunity, and cancer. A particularly appealing aspect of this approach is that a set of candidate advantageous SNPs associated with specific biological hypotheses were identified, which will be useful for follow-up testing. Jeong et al. [2018] combined approaches for detecting polygenic adaptations and for mapping the genetic bases of physiological and fertility phenotypes in approximately 1,000 ethnically Tibetan women from Nepal, who are adapted to high altitude living. The results of genome-wide association analyses and tests for polygenic adaptations showed evidence of positive selection for alleles associated with more pregnancies and live births and evidence of negative selection for those associated with higher offspring mortality. A lower hemoglobin level did not show clear evidence of polygenic adaptation, despite its strong association with an EPAS 1 haplotype carrying selective sweep signals.

Genomic analysis of high altitude populations residing in the Andes and Tibet has revealed several candidate loci for high altitude adaptation, a subset of which have also been shown to be associated with hemoglobin levels, including *EPAS1*, *EGLN1*, and *PPARA*, which play a role in the *HIF-1* pathway [Bigham et al. 2009; 2010; Simonson et al. 2010; Xu et al. 2011; Yi et al. 2010; Beall et al. 2010]. Scheinfeldt et al. [2012] highlight several candidate genes for high altitude adaptation in Ethiopia, including *CBARA1*, *VAV3*, *ARNT2*, and *THRB*. Although most of these genes have not been identified in previous studies of high altitude Tibetan or Andean population samples, two of these genes (*THRB* and *ARNT2*) play a role in the *HIF-1* pathway—a pathway implicated in previous work on Tibetan and Andean populations. The authors suggest that the genes and genetic variants contributing to high altitude adaptation in Ethiopia are largely distinct from those found in other high altitude regions and arose independently through convergent evolution due to the strong selective force of hypoxia. To date, dozens of genes have been found, which the authors believe may be related to the genetic adaptation of humans to hypobaric hypoxia. However, as can be seen from the above examples, much remains unclear.

So, do the results of these molecular biological (gene) studies meet the requirements of the modern theory of evolution? Our answer is probably more no than yes. Without going deeper into the theory, let us limit ourselves to the fact that, for all their importance, none of these studies have provided evidence of their selective value. In addition, there is no data answering the following questions: a) what are the phenotypic expressions of these adaptive genes?; b) when, where, and in whom did these adaptive genes arise?; c) what are the frequencies of these adaptive genes in individuals from different age groups in the population where such studies have been conducted?; d) are these genes are available in all high altitude human populations and if not why not?; e) whether such adaptive genes are found in animals permanently living at high altitudes?; and f) whether they are related to the development of pathologies that occur in high altitude conditions?

There are hundreds of definitions of adaptation in the literature. Ultimately, most authors agree that a trait is adaptive if it enhances the fitness of an organism, i.e., if the trait contributes to the survival and/or better reproductive success of an individual or social group. In many of his writings, E. Mayr rejected reductionism in evolutionary biology, arguing that evolutionary pressures act on the whole organism, not on single genes, and that genes can have different effects depending on what other genes represent. He rejected the idea of a gene-centered view of evolution, insisting “a gene is never visible to natural selection and in the genotype”. Mayr writes:

> “Evolution deals with phenotypes of individuals, with populations, with species; it is not a change in gene frequencies. … It is the phenotype that is exposed to natural selection, and not individual genes directly … Not its genes or genotype, because these are not visible to selection, but rather its phenotype. The word phenotype refers to the totality of morphological, physiological, biochemical, and behavioral characteristics of an individual by which it may differ from other individuals” [Mayr, 2002].

If genes are not the answer, then what is? Our own experience in the search for the genetic basis of human adaptation to extreme natural conditions in Eurasia (the Extreme North of Siberia; the high altitudes of the Pamirs and Tien-Shan) suggest that chromosomal Q-HRs constitute the relevant genetic material. That chromosomal Q-HRs meet the requirements of modern theory is based on the following facts: a) consistent interpopulation differences in the quantitative content of chromosomal Q-HRs in the genome of human populations have been established; b) these differences have been proven to be related to features of the ecological environment of the place of permanent residence, rather than to racial and ethnic composition; c) the quantity of chromosomal Q-HRs in a population's genome tends to decrease from low to high geographical latitudes, and from low to high altitude.

In addition to the above facts, which indicate the selective value of chromosomal Q-HRs, there are other data that speak in favor of our hypothesis: a) although chromosomal Q-HRs exist in the genome of only three higher primates (*H. sapiens*, *P. troglodytes*, and *G. gorilla*), their wide quantitative variability is solely a characteristic of human populations; b) the greatest number of chromosomal Q-HRs are present in the genome of gorillas and chimpanzees, and the smallest number in humans. It is notable that the orangutan has no such chromosomal segments; c) Q-HRs, in genetic terms, are completely inert, that is, these specific regions of chromosomes do not contain structural genes, and therefore their quantitative changes have no consequences for the informative part of the genome and can be carried out at an extremely high rate; d) individuals in human populations differ in the number, location, size, and fluorescence intensity of chromosomal Q-HRs in the genome; e) the results of extensive comparative cytogenetic population studies show that the populations of modern man are significantly different and these differences are associated with the natural environment of permanent residence, and not with racial or ethnic characteristics.

All these facts find their rational explanation in the hypothesis of cell thermoregulation. We have suggested the hypothesis of cell thermoregulation (CT), which was formulated based on a number of studies, primarily on the distribution of chromosomal Q-HRs in human populations.

In essence, the idea proposed can be reduced to the evolution of the genome structure and the physiology of the whole organism in higher eukaryotes in parallel to counteracting temperature changes in the ambient environment for more effective preservation of a constant temperature of the internal environment. The outcomes of such a parallel evolution were: (1) the appearance of different kinds of CC (C and Q-heterochromatin; G+ and Q+ bands), the effect of which at the genome level is generally subject to the laws of physics, and (2) the formation at the organism level of a complex organ-based (hypothalamus) physiological system of thermoregulation. This is why redundant DNA in the form of chromosomal HRs has no phenotypic expression and bears no specific function—

chromosomal HRs in CC participate in thermoregulation at the level of individual cells.

Human adaptation to cold and high altitude hypoxia, in general, can be described as follows. In the process of evolution, humans, chimpanzees, and gorillas inherited, among other things, chromosomal Q-heterochromatin from a common ancestor. However, this was subsequently distributed differently in their genomes, possibly as a result of random population genetic processes that took place in the early stages of divergence of these species. We can hardly hope to learn anything about the nature of these processes in the foreseeable future; however, we know that the ancestors of *H. sapiens* had Q-heterochromatin on seven autosomes and the Y chromosome; for *P. troglodytes* it is found on five autosomes only; and in *G. gorilla* it is found on eight autosomes and the Y chromosome [Pearson 1973, 1977; Seuanez et al. 1976]. Each of these species evolved independently.

However, it is possible that the unusual adaptive success of *H. sapiens* was achieved for the following reasons. The ecology of the Middle and Late Miocene was far from smooth and climatic changes, such as cooling, dryness, and seasonal and daily temperature fluctuations, gradually became the dominant environmental factors. Thus, our ancestors, perhaps even before leaving Africa, faced the problem of adaptation to new and more severe natural conditions, which were different from the climate of the African savannah.

According to our model of adaptive human evolution, it all began with the time when individuals with different quantities of Q-HRs (as happens now) began to be born in the populations of modern *H. sapiens* [Ibraimov 1993]. This was possible because: 1) the number of Q-polymorphic loci in the karyotype was big enough (25 loci) to allow for the appearance in human populations of individuals with different numbers of Q-HRs; 2) the relative frequencies of Q-HRs at these loci were different, so that the same number of Q-HRs in the genome in different individuals could be provided for by a variety of combinations of chromosomal Q-HRs [Ibraimov 2011, 2017b]. As such, we are dealing with a complex self-supporting genetic system, unique among classical genetic objects. This unique feature of chromosomal Q-HRs was taken advantage of by our ancestors with the onset of climate change in the savannah and especially when they went beyond it in search of new habitats.

It is hard to say why the ancestors of *P. troglodytes* and *G. gorilla* were unable to take this path. However, the following assumption can be made: the initial frequencies of chromosomal Q-HRs at all Q-polymorphic loci were high and individuals with different Q-HR numbers in the karyotype were born in the populations of these two higher primates. Therefore, the probability of occurrence of individuals with different conductivity, including those with low body heat, was simply excluded. In other words, chimpanzee and gorilla populations were initially unable to vary the number of chromosomal Q-HRs in the genome as widely as human populations [Ibraimov 1993, 2017b, 2018a]. The following facts favor this assumption: 1) the range of variability in the number of Q-HRs in the chimpanzee genome is from 5 to 7, whereas in human populations it is from 0 to 10; 2) in gorillas and chimpanzees, but not in humans, a special type of Q-heterochromatin, located at the distal ends of some chromosomes (7, 11, 20, and 23 in gorillas; 20, 21, 22, and 23 in chimpanzees), is found, which eliminates the possibility of birth individuals with different numbers of Q-HRs in populations.

How, then, does a human with relatively low body heat conductivity (BHC) adapt to high altitude climates? Obviously, chromosomal Q-HRs cannot directly affect the ability of Hb to bind molecular oxygen in the alveoli. However, we believe that low BHC helps man more effectively deal with hypoxia at high altitudes indirectly through heat energy and oxygen saving. The essence of heat saving is that individuals with low BHC are able to retain some of the metabolic heat in the body to maintain temperature homeostasis, which is usually lost to the environment through conduction, radiation, and evaporation. An increase in temperature reduces the possibility of binding hemoglobin oxygen. The temperature in tissues is higher than in the lungs. In this case, the warmer the body of the individual, the better the oxygen supply to his body will be due to the more facilitated dissociation of oxyhemoglobin, as a result of which the blood transfers the oxygen released from the chemical compound into the tissue liquid (oxygen saving). In speaking of energy saving, we mean that, to maintain a constant temperature in the body, individuals, even of comparable bodyweight, will require different amounts of energy. Consequently, individuals with low BHC will have some advantages at high altitudes—zones that are characterized by limited natural and food resources.

The views expressed here are best illustrated by the example of birds whose adaptability to the extremely low partial pressure of oxygen in atmospheric air is not in doubt. Of the two groups of homeotherms (birds and mammals), a higher core temperature (>41 °C) is found in birds. In birds, the structure of the respiratory organs differs significantly from mammals, i.e., the lungs of birds are small, with little flexibility, and they communicate with large thin-walled air bags that do not participate in gas exchange. According to their functional purpose, the air bags are divided into two groups—front and rear—through which air enters and leaves the respiratory system (lungs). In addition, these air bags are located in the abdominal cavity between the internal organs, some grow beyond this cavity between the muscles and, finally, inside the bones. This complex respiratory system provides a unidirectional flow of air—from the rear bags through the lungs to the front bags and then back out. During flight, there are no breathing movements of their own, and breathing occurs passively, due to periodic compression of the armpits and chest bags with each lowering of the wing; ventilation is enhanced as a result of powerful contractions of the chest muscles. Together, this kind of respiratory system means that the residual air in birds is limited mainly by air bags, where the air is heated to body temperature.

Traditionally, the adaptive value of air bags in birds is that they play the role of fur in mamals, blowing air through the lungs, as during flight the chest wall should be fixed to create sufficient support for the action of the wing muscles. It seems to us that the presence of a system of air bags in the deep parts of the body is not limited to the above effects. The adaptive value of air bags in hypobaric hypoxia seems to be much wider. It is to increase the temperature of the air in the body so as to create the best physical and chemical conditions for the dissociation of oxygen from hemoglobin. This may indicate that, in the genome of birds, for 130 million years of their existence on Earth, directed, favorable mutations, contributing to an increase in affinity Hb to O2 in hypobaric hypoxia, did not appear. Instead, birds developed an indirect way—facilitating the dissociation associated with Hb oxygen—by increasing blood temperature in the tissues.

In addition to low BHC, indigenous people deal with extreme high altitude climates through effective forms of behavior. For example, highlanders, unless absolutely necessary (for example, Sherpas accompanying climbers), rarely rise to great heights, as the daily life of the farmer takes place at an altitude below 5,000 meters above sea level. To overcome long distances or high terrain, highlanders use yaks or horses. In addition, the life of the highlander is closely related to the rhythm of nature; the way of life avoids performing heavy physical work and the sparing use of energy, often creating a false image of lazy or slow people among travelers.

Therefore, was *H. s. sapiens* able to adapt genetically to the influence of low temperatures different to those faced in East Africa? It is difficult to assume that in such a short time of existence at high altitudes, an adaptation for the effective supply of human cells with oxygen appeared and then successfully spread to the genomes of highlanders, directing favorable mutations in the genes involved in the long chain of oxygen transport from the alveoli to the tissues. This is indirectly evidenced by the known fact that places of permanent human settlement do not exceed 4,200 m above sea level.

Inevitably, the question arises as to when these hypothetical adaptive genes could appear? The problem is complicated by the issue of whether such genes (as in preadaptation) arose before man began to penetrate ecological areas that were extreme for a tropical species, or whether they appeared each time a human population entered and adapted to new conditions. If we recognize the possibility of the reemergence of specific adaptive genes in the human genome, do such reemergences apply to the same genes? Leaving open the question of the possibility of the existence of such highly mutable genes, let us ask another question: how did such favorable mutations spread in the genome of the inhabitants of high altitude populations? After all, high altitude populations are often isolated. The existence of some "high ceiling" (about 4,200 m above sea level) for permanent human habitation indirectly indicates the absence of specific adaptive genes in the genome of modern highlanders.

I, as a native of one of the most high altitude regions of the Eastern Pamirs, do not quite understand why hypobaric hypoxia, rather than cold, is singled out from among all the harmful physical factors of the environment found in the high altitudes? In addition to hypoxia, piercing cold, long winters, sharp seasonal and daily temperature fluctuations, dry air, frequent winds, ultraviolet radiation, and high terrain, etc. all have impacts on the human body at high altitudes. Furthermore, life at high altitudes, as nowhere else, requires high calorie food, warm homes, and thick clothing, which is not always available to all. All medical and biological studies have shown that permanently living at high altitudes does not contribute to good health, as evidenced by the high infant mortality rate and issues of reduced physical development, cold-related diseases (including seasonal colds and pneumonia), and malnutrition [Baker 1978].

In highlanders, chronic mountain diseases are not the most common pathologies and they do not have a significant impact on their demographic indicators. All chronic mountain diseases originate from the common cold. We tend to believe that the main harmful physical factor for human survival in the high altitudes is cold, rather than hypobaric hypoxia. After all, if there is not enough oxygen, a person can stop climbing and go down. But life in the conditions of high altitude cold with sharp seasonal and daily temperature changes is a constant struggle for existence, especially for a tropical species like man.

Unlike many animal species, man is unable to live in an environment of extreme cold. He is basically a tropical homeotherm. The reasons why exactly *H. sapiens* managed to master the entire landmass of the Earth, while remaining a single tropical species, is still unclear. We believe that it is highly probable that, in adapting to a climate so very different to that of East Africa, humans engaged a non-genic part of their genome known for its high mobility and nonconservatism in individual development and evolution. Moreover, we believe that man, as a species, owes its origin primarily to chromosomal Q-HRs, rather than specific structural genes [Ibraimov 1993, 2019b, e, f].

The problem of human adaptation is inextricably linked to one of origins. Below, we present our vision of the origin of modern man based on the wide polymorphism of chromosomal Q-HRs. The origin of modern humans is proposed on the basis of studies into chromosomal Q-HRs. This model is based on the following facts: a) chromosomal Q-HRs are found in the genome of only three higher primates (humans, chimpanzees, and gorillas); b) chromosomal Q-HRs in human populations, unlike those in apes, exhibit considerable quantitative variability; c) chromosomal Q-HRs participate in cell thermoregulation, which affects the maintenance of temperature homeostasis in the body; d) chromosomal Q-HRs display a physiological phenotype at the level of the whole organism, which can be objectively assessed; e) phenotypic manifestations of the variability of the quantity of chromosomal Q-HRs suggest a different level of body heat conductivity; and f) the quantity of chromosomal Q-HRs in the genome has a selective value in the adaptation of human populations to various environmental conditions.

In 1950, the evolutionist Ernst Mayr argued that the human phylogeny consisted of a mere three species in a single lineage. By the end of the twentieth century, the known hominid genealogical tree contained twelve species. The latest version of that tree covers twice as many species of which *H. s. sapiens* is the only hominid species remaining in the world today. The question arises: why, all other things being equal, was it the ancestors of modern man that managed to survive?

We do not wish to dwell particularly on current ideas about the origin of modern humans. All of them directly or indirectly recognize the primary role of genes. We merely wish to imagine how the distinctive biological features of contemporary man listed below could have originated. These characteristics include: a) only modern man has managed to master all the climatic and geographic provinces of the Earth, while remaining a single tropical biological species; b) humans display very high physiological plasticity; a) they have a large neocortex; d) they show an ability for conceptual thinking; e) they have mastered speech; f) they have hairless skin; g) they have 46, rather than 48 chromosomes; h) the broad quantitative variability of chromosomal Q-HRs is found only in human populations; i) we can identify the emergence of purely human pathologies (alcoholism, drug abuse, obesity, atherosclerosis, and high altitude pulmonary edema); and i) they have the ability to run long distances. Clearly, this is far from being a full list of characteristic traits of *H. s. sapiens*. The most remarkable thing is that all these exclusively human biological traits have appeared over a very short (in terms of evolution) period of time.

We believe that all these human features cannot be the product of directed favorable mutation of genes. The assumption of such a possibility would mean the existence of super mutable genes only in the human genome or the recognition of the existence of a goal in nature, neither of which correspond to scientific experience. And what do we offer in return? We believe that, in the genome of the higher primates, there is genetic material whose phenotype can be studied and used to rationally explain the origin of all exclusively human traits. This genetic material is constitutive heterochromatin. We examine below our objections to the role of genes in the origin of modern man.

First, we present those facts that do not favor genes as the source of human adaptation. Even before the Human Genome Project, there were doubts about the primary role of genes in the origin of modern man. For instance, comparison of the length of chromosomes in humans and chimpanzees showed the total amount of chromosomal material in these two species to be very similar; they differ only in the distribution of heterochromatin regions and in several pericentric inversions and translocations [Paris Conference 1971; Warburton et al. 1973; Yunis and Prakash 1982]. Experiments on DNA hybridization suggest that at least 98 % of nonrepeated DNA is identical in man and chimpanzee—a value compatible with the appearance of a viable hybrid. Summarizing numerous studies performed in different laboratories around the world, King and Wilson [1975] estimated that human and chimpanzee proteins are 99 % identical. Such considerable similarity makes it difficult to explain the great biological differences observed between these two closely related species. To date, no one has been able to name even a single gene that could be considered "responsible" for our origin.

So, for example, biochemists are now able to compare the DNA and protein sequences of modern animals and on the basis of the small differences found, they can estimate the chemical similarity of man and the lower primates. The similarity appears to be very high: DNA differs by about 2 % and proteins by about 0.3 % between humans and chimpanzees. Based on these small differences and on estimated rates of sequence variation over time, it has been deduced that a hypothetical last common ancestor to man and African apes existed as recently as five to seven million years ago [Sarich 1971; Sibley and Alquist 1984; Marks et al. 1988]. However, only a very tiny fraction of those fixed differences gave rise to the different phenotypes of humans and chimpanzees and discovering those is a great challenge. The vast majority of such differences are neutral and do not affect the phenotype. After the success of the Human Genome Project and the genome sequencing of the most studied prokaryotic and eukaryotic species, it became even more difficult to consider the primary role of genes in the origin of modern man.

In the following, we provide evidence of the possible selective value of chromosomal Q-HRs:

a) Consistent interpopulation differences in the quantitative content of chromosomal Q-HRs in the genome have been established [Buckton et al. 1976; Lubs et al. 1977; Al-Nassar et al. 1981; Ibraimov and Mirrakhimov 1982 a, b, c, 1985; Ibraimov et al. 1982, 1986, 1990, 1991, 1997, 2013; Stanyon et al. 1988; Kalz et al. 2005; Decsey et al.].
b) These differences have proven to be related to features of the ecological environment of permanent residence, and not to racial and ethnic composition [Ibraimov and Mirrakhimov 1985; Ibraimov 1993, 2015, 2017].
c) The quantity of chromosomal Q-HRs in a population's genome tends to decrease from low to high geographical latitudes and from low to high altitudes [Ibraimov and Mirrakhimov 1982 a, b, c, 1985; Ibraimov et al. 1982, 1986, 1990, 1991, 1997, 2013].
d) The Q-HR block on the Y chromosome is the largest in the human karyotype, with an average size twice that of all the Q-HRs on the autosomes taken together. The size of the Q-HR segment on the Y chromosome influences the quantities of Q-HRs on the autosomes. For example, in males with large blocks of Q-heterochromatin on the Y chromosome, the number of Q-HRs on their autosomes is lower and *vice versa* [Ibraimov et al. 2000; Ibraimov 2014].
e) The overall quantity of Q-HRs on the autosomes in females is higher than in males. The increased number of chromosomal Q-HRs on autosomes in females at the population level can be explained by the existence of some evolutionarily established mechanism that "compensates" for the difference in the "dose" of

Q-heterochromatin material in the female genome due to the lack of a Y chromosome in the female karyotype [Ibraimov 2014].

f) Different age groups have different quantities of chromosomal Q-HRs: the greatest number of Q-HRs is found in neonates, while the smallest is found in the elderly [Buckton et al. 1976; Ibraimov and Karagulova 2006a; Ibraimov et al. 2014a].
g) In the first few years of life, *ceteris paribus*, infants that die among healthy children often have the greatest number of Q-HRs in their genome [Ibraimov and Karagulova 2006b].
h) Individuals capable of successfully adapting to extreme high altitude climates (e.g. mountaineers) and the Far North (e.g. oil workers on the Yamal peninsula of Eastern Siberia) are characterized by extremely low quantities of Q-HRs in their genome [Ibraimov et al. 1986, 1990, 1991].
i) High altitude pulmonary edema can develop in an individual who has a large number of chromosomal HRs in his genome [Ibraimov 2018].
j) All forms of purely human pathology (alcoholism, drug addiction, obesity, and atherosclerosis) are associated with a wide quantitative variability of chromosomal Q-HRs. For example, individuals with a smaller quantity of Q-HRs in their genome are prone to alcoholism and obesity, while those with a greater quantity of Q-HRs are prone to drug addiction [Ibraimov 2016a, b, c, 2017c, 2018a, b, c].
k) Finally, unlike hypothetical adaptive genes, the quantity of chromosomal Q-HRs in the human genome has a clear physiological phenotype expressed in the form of variable body heat conductivity [Ibraimov et al. 2014b].

That the amount of chromosomal HRs in the human genome may have selective value, we explain through the hypothesis of cell thermoregulation [Ibraimov 2003, 2004, 2017]. We have suggested the CT hypothesis, formulated through our studies, as relying mainly on the distribution of chromosomal Q-HRs in human populations. We suggest that condensed chromatin (CC), which includes chromosomal Q-HRs of higher eukaryotes, is likely to relate to the thermoregulation of cells. CC, being the most densely packed material, appears to have the greatest heat conductivity in the interphase cell [Ibraimov 2003, 2017, 2019e, f].

We have checked this hypothesis at the level of the human organism assuming that CT is the basis for heat conductivity of the whole cellular part of the body. The hypothesis that the quantity of Q-HRs in the genome is likely connected to the wide variability of human body heat conductivity has been proved. The results obtained show that individuals in a population truly differ from each other in terms of BHC and its level depends on the quantity of chromosomal Q-HRs in the genome (for more details see Ibraimov et al. 2014b).

Now, let us consider in more detail whether our hypothesis is able to shed light on the origin of modern man. This possibility we consider in terms of the influence of chromosomal HRs on the appearance of important and distinctive human features. These features include: a) the ability to adapt to the different climatic conditions of the Earth; b) high physiological plasticity; c) a large neocortex; d) the ability to engage in conceptual thinking; e) a capacity for speech; f) hairless skin; g) 46 rather than 48 chromosomes; h) the wide quantitative variability of chromosomal Q-HRs; i) the existence of purely human forms of pathology; and j) the ability to run long distances (see above).

Our model of adaptive human evolution suggests that it all began at the time when individuals with different amounts of Q-heterochromatin material began to give birth in various populations (as happens now). This unique feature of chromosomal Q-HRs was taken advantage of by our ancestors with the onset of climate change in the savannah and especially when they sought to go beyond its limits in search of new habitats [Ibraimov 1993].

It is generally considered that humans are well adapted to hot climates. This is connected to the fact that modern humans are a biological species that developed in the tropical climate of East Africa. Unlike many animal species, man is unable to live in an extremely cold environment—he is basically a tropical homeotherm. However, for various reasons, human populations live under conditions of low or high environmental temperature (*T*) where maintaining temperature homeostasis is especially difficult. Three factors effect thermo-regulating systems: heat production, heat loss, and thermoregulatory behavior. Though important, they cannot be effective from a long-term perspective.

We suppose that *H. sapiens*, as with all mammals, possesses an additional but appropriate and simple mechanism of thermoregulation. In the present case, in order to preserve temperature homeostasis under different environmental conditions, in addition to physiological, behavioral, and biochemical mechanisms, such as wide intrapopulation variability, BHC came into play [Ibraimov 1993, 2003, 2004; Ibraimov and Tabaldiev 2007]. It is possible that BHC diversity is necessary for *H. sapiens* because no single genotype can possess sufficient adaptadness for all environments.

On the whole, we see the maintenance of temperature homeostasis under conditions different from those of East Africa as follows: 1) individuals with fewer chromosomal Q-HRs in the high geographic latitudes more effectively maintain temperature homeostasis because of their low BHC, permitting the preservation of an additional amount of produced heat and slowing down the body's cooling rate due to external cold; 2) individuals with high BHC in high geographic latitudes, constantly lose additional metabolic heat through conduction, exposing them to relatively fast cooling due to cold. As a result, such individuals have to produce more heat and/or consume more high-calorie food for heat production, which is not always simple or healthy; 3) individuals with low BHC in low geographic latitudes (where environmental temperatures are often higher than body temperature) receive additional heat from the environment by means of conduction, which is not used in useful physiological work. That is why these individuals overheat faster and have to transfer the heat surplus (through sweating, polypnoea, forced rest, behavioral reactions etc.) to the environment at the cost of a significant decrease in physical and mental activity, negatively influencing their adaptation to hot climates; finally, 4) individuals with a large number of Q-HRs in the genome at low latitudes, whose bodies have high thermal conductivity, are perhaps able to better adapt to high environmental temperatures and more effectively level off temperature differences across the body through a rapid transfer of surplus heat from the organism to the environment, including heat radiation.

Generally speaking, we wish to state that individuals with high BHC better and more rapidly equalize temperature differences across the body and *vice versa*. If this is true, then it provides a simple explanation, for example, for the known resistance of southern natives to high environmental temperatures. This means that they are able to effectively equalize temperature differences in different parts of the body and more rapidly direct surplus heat to the environment. In contrast, indigenous peoples living in the Far North or at high altitudes are better, and for longer, able to maintain metabolic heat in the body due to the low heat conductivity of their bodies, with all ensuing consequences. This might also explain why males better endure heat than females, while the latter have a more stable response to cold than males.

Taking into consideration all the above, we can explain why the quantity of chromosomal Q-HRs is greater in the genomes of newborns, than in senior age groups [Buckton et al. 1976; Ibraimov and Karagulova 2006a], and why the same chromosomal material is found in greater quantities in the genomes of infants who died during the first few years of life [Ibraimov and Karagulova 2006b]. The prevalence of people with a smaller quantity of chromosomal Q-HRs in the genome in senior age groups may be connected to the negative selection of individuals with a greater quantity of Q-HRs during the first years of life. The infant ratio of body surface to body capacity is higher than the adult ratio. When one more physical factor (high BHC) is added to this, then these infants are more vulnerable to colds and their consequences.

The rapid (on an evolutionary scale) and effective mastering of the total oikumene by man is indeed a unique phenomenon and this makes one ponder the possibility that structural genes are not involved, but rather some mobile, non-conservative part of the genome. Thus, our data suggests that *H. sapiens* used the Q-heterochromatin part of the genome for adaptation to the cold [Ibraimov et al. 1982, 1986, 1990, 1991, 1997; Ibraimov and Mirrakhimov 1982 a, b, c, 1985]. However, we can hardly imagine as yet how this genetically inert material could be used in this adaptation. Therefore, in order to relate a possible mechanism through which chromosomal Q-HRs can influence human activity, including permanent residence in hot and cold climates, we are compelled to admit, first of all, that chromosomal Q-HRs as condensed chromatin have a heat conductive effect in cells [Ibraimov 2003, 2004, 2015a, b, 2017].

We suppose that *H. sapiens*, besides those mechanisms inherent in other mammals, possesses an additional, but appropriate and simple mechanism of thermoregulation. In the present case, in order to preserve temperature homeostasis under different environmental conditions, and in addition to various physiological, behavioral, and biochemical mechanisms, the wide variability of BHC operates [Ibraimov 1993, 2003]. For *H. sapiens*, it is possible that diversity of BHC is necessary because no single genotype can possess sufficient adaptability for all environments.

It is generally considered that *H. sapiens* is characterized by high physiological plasticity. It appears likely that when speaking about this plasticity, we refer to the ability of man to inhabit very different climatogeographical regions. However, we also assert that the basis of the high physiological plasticity of *H. sapiens* is due to the wide quantitative variability of chromosomal Q-HRs in human populations, which, through changes in the physical density of condensed chromatin in the cells, exert a modifying influence on the level of heat conductivity of the human body as a whole.

In reality, one and the same person cannot possess equally good adaptation to heat, cold, and high altitude hypoxia. However, in any human population there are individuals able to efficiently adapt to tropical climates, the Far North, and high altitude conditions. It should be understood that man, as a biological species, but not as an individual, can adapt to heat, cold, or high altitude hypobaric hypoxia.

We assume it is possible that, with such high physiological plasticity in *H. sapiens*, there is wide variability of BHC in human populations. Selection does not affect single individuals, but the local population [Mayr 1970]. As such, we believe that: (1) human bodies in a population significantly differ from each other in terms of heat conductivity and (2) organ-based physiological thermoregulation in man is realized in different physical conditions in the form of different BHCs.

The example of modern sport can help illustrate our understanding of the role of chromosomal Q-HRs. More and more countries situated at low geographic latitudes have started taking part in competitive world sport. Notably, competitors residing in these regions are achieving great success in sports requiring (in addition to other factors) effective heat-loss (such as football, boxing, and marathon running). In contrast, sportsmen from higher latitudes prevail in water and winter sports, and also in mountaineering [Ibraimov 2019c; Ibraimov et al. 1990, 1991]. It has been ascertained that natives of low latitudes have more chromosomal Q-HRs in their genome [Lubs et al. 1977; Ibraimov and Mirrakhimov 1982 b, c, d, 1985; Ibraimov et al. 1997; Kalz et al. 2005]. As we think, since the bodies of southerners have relatively high BHC [Ibraimov and Tabaldiev 2007], it is not surprising that they are successful in sports that require effective heat loss. Indeed, a sportsman with high BHC cannot make much progress in water sports due to the fact that his body will cool too rapidly. However, this sportsman is likely to be more successful in sports requiring effective heat loss. Looking at these examples, it is not difficult to imagine man's capacity for long distance running.

Comparative tests of the endurance of "whites" and "blacks" to physical loads in conditions of heat and high humidity demonstrate the superiority of Negroid individuals even over Europoids who are used to working in such conditions. On the other hand, the experience of the Korean War shows that frostbite occurs much more often in Negroid soldiers than in Europoids (Folk 1974). The long-term experience of Indian medical officers in the Himalayas has shown that, under identical environmental conditions, southern Indians are physiologically more susceptible to frostbite and cold injures, while high altitude populations, such as Gurkhas and northern Indians, are more resistant (Mathew 1992).

On the question of the origin of the large neocortex and hairless skin in humans, we have devoted a special study [Ibraimov 2007], the essence of which is as follows. The plausible reason for this phenomenon is that skin, after having lost its hair, became the largest sense organ, which begins functioning as early as the prenatal period of human development. It is assumed that *H. sapiens* became the possessor of the largest brain among the primates after losing the hair on its skin. It appears that human skin became hairless due to the evolution of condensed chromatin in the genome of the direct ancestors of modern humans. In particular, in their karyotypes together with chromosomal Q-HRs, three pairs of autosomes (1, 9, and 16) and the Y chromosome became the carriers of unusually large C-HRs, leading to a significant increase in body heat conductivity. In the conditions of tropical Africa, inhabited by our ancestors, the variability of skin covered with hair became a serious obstacle to maintaining temperature homeostasis, particularly in terms of dissipating excess heat from the organism, which finally led to hairlessness. Given this fact, the heat protective function of hair cover was taken over by a large amount of subcutaneous adipose tissue.

The wide quantitative variability of chromosomal Q-HRs in humans, in our opinion, is associated with the origin of 46, rather than 48 chromosomes, in the human genome [Ibraimov 2017b]. It is an established fact that the number of chromosomes in chimpanzees, gorillas, and orangutans is 48. However, the cause and effect of such chromosomal rearrangement is unknown. It has been proposed that natural selection caused the merger of two pairs of autosomes into a single chromosome. In the changing climate of East Africa, individuals with a smaller quantity of chromosomal Q-HRs in the genome were best adapted to the environmental conditions. Two pairs of acrocentrics in the genome of the common ancestor, which merged into a single chromosome, appear to have carried the short arms of Q-HRs with a very high frequency and prevented the birth of individuals with a small quantity of Q-heterochromatin. With the merger of these two pairs of acrocentrics into one, the number of autosomes bearing Q-HRs reduced from nine to seven pairs, as is the case in modern humans. Such chromosome rearrangement resulted in two important consequences: a) chromosomal Q-HRs were distributed across seven Q-polymorphic autosomes, so that it was possible to give birth to individuals with different, including small, numbers of Q-heterochromatin; b) in human populations, individuals with low numbers of Q-HRs appeared, who were able to adapt to new and harsher climatic conditions. Over time, these individuals formed new populations in new territories, where individuals with a similar number of chromosomal Q-HRs to modern natives of Africa, and with 46 chromosomes in the genome, began to dominate. Thus, the origin of the 46 chromosome karyotype, from an ancestral 48 chromosome line, was natural selection and its effect was adaptation, i.e., individuals with different, including low, numbers of Q-HRs, gained the advantage of being able to open up and colonize new ecological zones in East Africa and beyond (for details, see Ibraimov 2017b).

The existence of purely human pathologies is well known. However, the reason for their occurrence in humans, it seems, has not been specifically considered. Our experience of studying the wide quantitative variability of chromosomal Q-HRs in such pathologies shows that it has some relation to the pathogenesis of obesity, alcoholism, drug addiction, and atherosclerosis, as well as high altitude pulmonary edeme (HAPE). A study established that patients differed significantly in terms of low BHC when compared to a control sample. Among such patients, those with drug addiction and HAPE had the highest BHC, followed by alcoholics and individuals suffering from alimentary obesity [Ibraimov 2016a, b; Ibraimov et al. 2010a, b, 2018a].

The possible role of BHC in these situations seem to be as follows: the frequency of taking strong alcoholic drinks displays an increasing trend from low to high latitudes and from low to high altitude, whereas the quantity of Q-HRs in the genome tends to decrease with the geographical latitude and altitude of the permanent residence of a population [Ibraimov 1993; Ibraimov and Mirrakhimov 1985; Ibraimov et al. 1982a, b, c, 1982, 1986]. Let us present a suitable example. In a sense, living in the Far North or at high altitude may predispose individuals to take strong drink for thermal comfort. However, given this, we assume that one and the same dose of alcohol taken by persons with different BHCs may result in different consequences. As such, in an individual with low BHC, alcohol intoxication begins after taking a relatively large amount of alcohol because of the lower temperature leveling in different parts of body, finally leading to stronger intoxication and a hangover, than with persons with normal or high BHCs. In other words, the lower the BHC of an individual, the slower the intoxication begins. This is due to the shorter time needed to heat the whole body and develop a sense of thermal comfort in the whole organism.

In individuals with high BHC (because of a large quantity of Q-HRs in the genome), drug addiction also appears due to a desire for rapid thermal comfort, but this "pleasure" is a result of a "drug overcooling" of the body with subsequent emotional or other experiences. We believe that the psycho-emotional effects of alcohol and drugs on the organism can be attributed to the degree of violation of temperature homeostasis, manifested in quite opposite directions. In other words, while ethanol causes alcohol intoxication and increases the body's temperature (oxidation of 1 gram of ethanol produces 7 kilocalories), drugs, on the contrary, lower the body's temperature, thus causing a state of drug-induced stupor. Nevertheless, the notorious susceptibility of southerners to drugs and northerners and highlanders to strong alcoholic beverages may be explained, to a certain degree, by the different Q-HR content of their genomes [Lubs et al. 1977; Ibraimov and Mirrakhimov 1982a, b, c; Ibraimov et al. 1990, 1991, 1997] and, accordingly, be related to human BHC.

As we have shown, individuals with lower quantities of Q-HRs in the genome are more prone to developing alimentary obesity [Ibraimov 2016b]. The results of numerous epidemiological studies, which have been carried out in many countries and regions, have unambiguously shown that females suffer from obesity twice as often as males. We assume that alimentary obesity is not the result of lacking inner discipline in food consumption or the presence of a hypothetical "gene for obesity". In individuals with low BHC, which is characteristic of the female organism in general, even if a normal amount of food is taken, more fat will be stored in the body than in persons with a high BHC. It is easy to imagine that individuals with a good heat isolating body, when consuming food rich in calories that is easily assimilated, and who are living in conditions of our contemporary sedentary life, become more vulnerable to developing alimentary obesity [Ibraimov 2016b].

We suggest that infants with a greater quantity of chromosomal Q-HRs in their genome are more frequently subject to over-cooling and catarrh etc. due to the high heat conductivity of their bodies. In contrast, individuals with a low BHC may have a certain advantage concerning their survival through infancy compared to those who have a medium or especially large quantity of chromosomal Q-HRs in the genome. This is how we explain the "redundancy" of individuals with a smaller quantity of chromosomal Q-HRs in the genome of elderly subjects [Ibraimov and Karagulova 2006a; Ibraimov et al. 2010a, b].

We suggest that atherosclerosis is a purely human disease that appeared after the adaptation of man to the climatic conditions of temperate and high latitudes in the northern hemisphere. Our hypothesis is based on the following prerequisites: 1) man was, and remains, the only tropical biological species whose ancestors, with the exception of the last 30-50,000 years, have lived permanently under tropical climate conditions; 2) atherosclerosis came into existence only after man was able to leave the boundaries of his original land (tropical East Africa); 3) the atherosclerotic process can develop into a pathology under conditions different from the climate of tropics, i.e., where humans breath cold air during a long period of the day or year; 4) the frequency and severity of the atherosclerotic process increase as geographical latitude and altitude above sea-level increase; 5) the possible cause of the deposition of "fatty streaks" in vascular walls is the blood cooling after pulmonary circulation; 6) the absence of atherosclerotic changes in the body's circulation, including the pulmonary artery, is due to the fact that warm (37.0 °C) blood always circulates in normal states; 7) the "atherogenic" effects of cooled blood should be, for thermophysical and hydrodynamic reasons, most pronounced in those vascular areas where turbulent rather than laminar blood flow occurs; in other words, where there is bifurcation and branching, as well as in the heart where the contraction of ventricular muscles also creates regular squeezing of the coronary vessel lumen [Ibraimov 2008]. The primary causative factor in the development of atherosclerosis is blood temperature. The degree of blood cooling in the lungs depends on the geographical latitude and altitude above sea level of the site of permanent residence. The preclinical stage of atherosclerosis may develop into a pathological form predominantly in individuals whose genome sees a higher than mean value of chromosomal Q-heterochromatin material per individual in the population [Ibraimov 2017c].

High altitude pulmonary edema (HAPE) occurs in unacclimatized individuals who are rapidly exposed to altitude in excess of 2,500 m above sea level. A working hypothesis on the etiology of HAPE suggests that hypoxic pulmonary vasoconstriction is extensive and precapillary resistance is elevated. This results in dilatation of the capillaries and capillary injury, with leakage of protein and red cells into the alveoli and airways. However, the question remains: why does HAPE develop only in some individuals when rapidly exposed to high altitude?

Over the last 50 years, scientists at our center have systematically studied the inhabitants of the high altitude areas of the Pamir Moutnains and the Tien-Shan. During these years, we have observed only one case of HAPE occurring in a colleague (male, 24 years old, and physically healthy) in the Eastern Pamirs at an altitude of 3,600 m above sea level. Everything ended well after the plane landed at an airport at a lower altitude. After returning to the center, all members of the expedition examined their karyotypes, including the polymorphism of chromosomal Q-HRs. It turned out that they were not different from normal individuals in terms of the number, location, size, and fluorescence intensity of chromosomal Q-HRs. The only difference was in the single individual who developed HAPE who found a large number of chromosomal Q-HRs in his karyotype (Fig. 3.2.5.1). Therefore, it is possible that, all other conditions being equal, HAPE most often develops in individuals with a large quantity of chromosomal HRs in the genome (for more details, see Ibraimov 2018c).

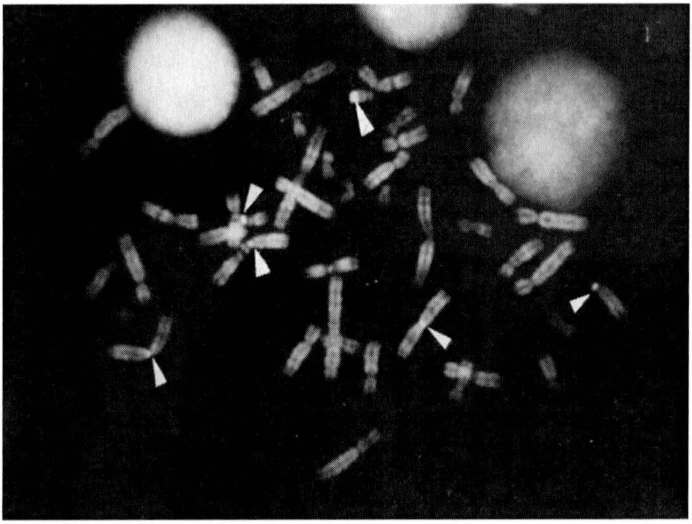

Fig. 3.2.5.1. Q-heterochromatin regions localized on the Q-polymorphic loci of three autosomes (3cen, 3cen, 13p11, 13p11, 13p13, and 21p13) and on the q12 locus of chromosome Y.

One can only speculate about the emergence of conceptual thinking and speech in humans [Ibraimov 1993, 2007, 2010, 2019b]. The wide quantitative variability of chromosomal Q-HRs in the populations of our ancestors allowed the emergence of individuals with different BHCs, providing an effective mechanism to survive in new and more severe climates than that of East Africa. Perhaps these events proceeded according to the following scheme. The ancestors of gorillas, chimpanzees, and humans all saw the emergence of Q-HRs distributed on their chromosomes. *H. sapiens* began to bare individuals with different distributions, including those with a small number of Q-HRs in the genome (preadapted) ⊠ *H. sapiens* expanded its habitat ⊠ in these populations, a range of idioadaptive changes in morphology and behavior emerged ⊠ these hominids began to accumulate collective experience and the need for its preservation and transmission saw the development of conceptual thought and speech ⊠ the fate of the species mainly determined its cultural and social, rather than biological, inheritance [Ibraimov 1993, 2007, 2017b].

Data analysis of the distribution of chromosomal Q-HRs in human populations has convinced us that the distributions of these Q-HRs on seven potentially Q-polymorphic autosomes are far from accidental. At the level of human populations, each of the seven Q-polymorphic autosomes contains a comparable "portion" of the overall quantity of Q-HRs in the genome, irrespective of the peculiarities of race and ethnicity. As such, it has become obvious that interpopulation heterogeneity formed due to a proportional increase or decrease in the absolute number of Q-HRs simultaneously in all potentially Q-polymorphic loci. Moreover, the data show that modern humans originated, most probably, from a single population limited in number. Speaking of the merit of the issue discussed—the origin of modern humans—the cytogenetic data presented above support the Out of Africa Model [Ibraimov 1993, 2011a]. Finally, the phenomenon of wide quantitative variability in chromosomal Q-HRs could shed light on the disappearance of Neanderthals. Neanderthals (*Homo neanderthalensis*) were widespread across Europe and Western Asia for a long time, starting from about 400,000 years ago. Things began to change when populations of *Homo sapiens* (earlier members of our own species) migrated from Africa to Europe about 45,000 years ago. 5,000 years later, not a single Neanderthal remained.

Over recent decades, two main theories on their demise have emerged. The first concerns climate change—their decline coincides with a period of extreme cold in Western Europe, which would have placed a huge amount of stress on the species. The other concerns competition with modern humans—our bigger brains and better adaptation to the environment at the time may have meant that the Neanderthals did not stand a chance. In terms of chromosomal Q-HRs, it can be assumed that the Neanderthals disappeared due to their inability to give birth to children with different, including low, quantities of Q-HRs in the genome. Chromosomal Q-HRs in the Q-polymorphic loci of Neanderthals have a different distribution pattern to those found in humans—their distribution was possibly uniform and very high in frequency in all loci, precluding the appearance of individuals with low quantities of Q-HRs in the genome. The inability to give birth to children with a low quantity of chromosomal Q-HRs, and with a low BHC, would have dramatically reduced their chances of survival in conditions of extreme cold. The emergence of a large brain volume in modern man may also be related to the quantity of chromosomal HRs, as has previously been discussed [Ibraimov 2007]. The superiority of man over Neanderthal may not have been down to brain volume, but rather to humanity's accumulated cultural and social experience in the course of adapting to the different ecological conditions of East Africa and Eurasia.

3.3. Concluding remarks

In an attempt to understand the biological role of noncoding DNA (ncDNA) in general, and chromosomal HRs in particular, we consciously abandoned the molecular biology approach. By no means diminishing the value of research aimed at elucidating the molecular peculiarities of the heterochromatic part of the genome, we are still inclined to believe that heterochromatin as studied by molecular biologists and the heterochromatic regions of chromosomes observed under a microscope are not the same.The fact is that it seems to many people that the problem of variability of chromosomal HRs in human populations can be reduced to the molecular biology of heterochromatin. However, in the last century this tendency has gradually been revised. As an illustration, I will cite here the statements of one of the leading specialists in the molecular biology of heterochromatin. Beridze [1982], the author of the first monograph on satellite DNA, writes:

> "In this matter, two aspects should be clearly distinguished: the first is the compaction of satellite chromatin and the formation of a condensed state, known as 'constitutive heterochromatin', the second is a function of the most constitutive heterochromatin, structure, which is a supramolecular formation, functioning as a qualitatively new unit. With the formation of constitutive heterochromatin, some new properties are acquired that are not characteristic of either satDNA or proteins that are part of the constitutive heterochromatin separately, that is, the properties of constitutive heterochromatin are not the sum of the properties of its components. ... The functions of constitutive heterochromatin should be considered separately. At the moment, the question of the functions of heterochromatin is a cytogenetic problem rather than a molecular biological one".

Q-HRs, in our view, constitute special chromosome segments representing a single self-sustaining structural-functional system, of which the quantitative content in the genome is controlled by natural selection. The effects of chromosomal Q-HRs on the human body depend on their total number in the genome and not on their localization on a particular chromosome in the karyotype. Q-HRs are unique chromosome regions only found in the genome of the three higher primates and their wide quantitative variability is inherent only in human populations [Pearson 1973, 1977; Seuanez et al. 1976; ISCN 1978; Geraedts and Pearson 1974; Müller et al. 1975; McKenzie and Lubs 1976; Buckton et al. 1976; Lubs et al. 1977; Yamada and Hasegawa 1978; Al-Nassar et al. 1981; Ibraimov and Mirrakhimov 1982 a, b, c, 1985; Ibraimov et al. 1982, 1986, 1990, 1991, 1997, 2013; Stanyon et al. 1988; Kalz et al. 2005; Decsey et al.

2006]. Data on the mean numbers and distribution of Q-HRs in natural populations of chimpanzees and gorillas are not available. However, the literature suggests that the largest number of Q-HRs is found in the genome of gorillas and chimpanzees, and the smallest number is found in humans. It is notable that in the orangutan, such brightly fluorescent chromosomal segments after differential staining are absent [Seuanez et al. 1976].

Studies of the patterns of chromosomal Q-HR distributions speak in favor of our view that Q-HRs at the level of human populations behave as a single self-sustaining genetic system whose behavior can be predicted. For example, when the mean values of Q-HRs increase, the absolute frequencies of Q-HRs usually increase on all seven Q-polymorphic autosomes in a population and *vice versa* [Ibraimov 1993, 2011a]. Populations with low mean values of Q-HRs per individual have a narrow range of variability in the numbers of Q-HRs in a population and *vice versa* [Ibraimov et al. 1986, 1990, 1991].

Regarding our hypothesis that the quantitative content of Q-heterochromatin in the genome is under the control of natural selection, data highlight the relationship between the quantity of chromosomal Q-HRs and the natural environment of the population's permanent residence. In particular, the content of Q-HRs is reliably lower in the genome of populations permanently residing at high latitudes and high altitudes, as well as among alien individuals well adapted to the extreme natural conditions of high altitudes (mountaineers) and the Far North (drilling workers in the oil fields of the Arctic Circle), compared to populations living in the temperate zones of Eurasia and lowland subequatorial Africa [Ibraimov and Mirrakhimov 1985; Ibraimov et al. 1990, 1991].

The fact that the effects of chromosomal Q-HRs on the human body depend on their total number in the genome, and not on the localization of a particular chromosome in the karyotype, is indicated by the results of our research on the variability of body heat conductivity in normal humans and in those with some purely human pathologies (obesity, alcoholism, drug addiction, atherosclerosis, and HAPE).

The totality of existing data on human chromosomal Q-HRs suggest that it is no longer possible to consider them as selectively neutral genome structures, as they have a certain selective value, and their number in the genome is controlled by natural selection at all stages of human development. This is evidenced by the following data. The mean number of chromosomal Q-HRs per individual in a population is greatest among newborns than people in older age groups [Buckton et al. 1976; Ibraimov and Karagulova 2006a; Ibraimov et al. 2014a] despite the fact that the number, location, and size of Q-HRs in ontogenesis do not change [Phillips 1977; van Dyke et al. 1977; Robinson et al. 1977; McCracken et al. 1978]. This is apparently due to the fact that some children with more than the mean number of Q-HRs in the genome in populations may undergo negative selection even in the first years of life [Ibraimov and Karagulova 2006b].

It can be considered firmly established that there are significant differences in the quantitative content of chromosomal Q-HRs among individuals in human populations. Since it is known that both the complete absence and the maximum number of Q-HRs in the genome do not show any visible pathological or other phenotypic manifestations, what does this mean for a particular individual? We argue that: a) chromosomal Q-HRs in the composition of condensed chromatin participate in cell thermoregulation and through it determine the level of heat conductivity of the cellular part of the human body; b) the number of chromosomal Q-HRs in the human genome may affect its ability to adapt to cold, heat, and hypobaric hypoxia; c) the number of chromosomal Q-HRs in the human genome may affect its vulnerability to some diseases of civilization, such as obesity, alcoholism, and drug addiction; d) the proposed effects of Q-HRs on the human body depend on their total number in the genome and not on their localization on a particular chromosome in the karyotype, i.e., they are not locus specific genetic structures.

We believe that *H. s. sapiens* is not bound by the presence of any specific purely human gene or complex of genes in the genome, but by the unique combination of two types of constitutive heterochromatin: C and Q. Speaking of C-HRs, we mean their total quantity, but not their localization; among the higher primates it is only in the karyotype of modern man that there are large blocks of C-heterochromatin (chromosomes 1, 9, and 16). It is the amount of this type of constitutive heterochromatin that man owes to the high level of his BHC. Although BHC has not yet been studied among higher primates, or in other homeothermic animals, we believe that it will be the highest of all in endothermic organisms. However, man owes a lot to his chromosomal Q-HRs. It is the wide quantitative variability of chromosomal Q-HRs in human populations that has ensured the place that modern man occupies. Thus, the bulk of existing data on human chromosomal Q-HRs testify to a self-sustaining genetic system subject to natural selection, and together with C-HRs they take part in cell thermoregulation, depending on the amount in the genome of a given individual, with all the ensuing consequences.

In conclusion, I would like to specifically emphasize one very important circumstance. Some readers may have gained the impression that the main role of condensed chromatin is cell thermoregulation. Of course, this is far from the case and we should remember this. Noncoding DNA make higher forms of supramolecular organization, nucleosomes, mitotic chromosomes, cell nuclei, all the condensed chromatin components (G+ and Q+ segments, chromosomal C-HRs and Q-HRs), cell thermoregulation, sex, multicellularity and homeothermic organisms, including us, possible. All these products of the evolution of noncoding DNA providing convincing examples of how the physical state of DNA molecules affect the vital activity of individual cells and the organism as a whole [Ibraimov 1993, 2003, 2004, 2007, 2008a,b, 2011a,b, 2012, 2015a,b, 2019g,h, 2020a,b,c].

The older generation of cytogeneticists, writing more than 50 years ago, were right:

> "… it has taken 100 years to achieve our current level of understanding of the nature and function of loci governing Mendelizing characters. The cytogenetics of the future must deal with that portion of the chromatin which is even more elusive as to its biochemical, developmental, and evolutionary roles" [Swanson et al. 1967].

They had heterochromatin in mind and, indeed, we are still faced with this reality today.

BIBLIOGRAPHY

Adkisson KP, Perreault WJ, Gay H. 1971. Differential fluorescent staining of Drosophila chromosomes with quinacrine mustard. Chromosoma. 34(2):190-205.

Allderdice PW. 1973. Identification of the location of a crossover in a pericentric inversion heterozygote which resulted in a duplication deficient chromosome 3. Am J Hum Genet. 25:11.

Al-Nassar KE, Palmer CG, Connealy PM, Pao-Lo Yu. 1981. The genetic structure of the Kuwaiti population. II. The distribution of Q-band chromosomal heteromorphisms. Hum Genet. 57:423-427.

Ami D, Di Segni M, Forcella M, et al. 2014. Role of water in chromosome spreading and swelling induced by acetic acid treatment: a FTIR spectroscopy study. Europ J Histochem. 58:2330.

Andrews P, van Couvering AH. 1975. In: Approaches to Primate Paleobiology. F.S. Szalay. (Ed). Karger, Basel, pp. 62-105.

Angell R. 1973. The chromosomes of Australian Aborigines. In: The human biology of Aborigines in Cape York J. Kirk (Ed). Canberra: Australian Institute of Aboriginal Studies 103.

Armendares S, Lisker R, Mutchinick O, et al. 1983. Absence of correlation between Y chromosome heterochromatin and several anthropometric measurements in a Mexican population. Hum Genet. 63: 283.

Arrighi FE, Hsu TC. 1971. Localization of heterochromatin in human chromosomes. Cytogenetics. 10:81-86.

Aula P, Saksela E. 1972. Banding characteristics of paracentric marker constrictions in human chromosomes. Hereditas. 70:309-310.

Autkiewicz HL. 1987. Chinese alcohol use and Hong Kong. Alcoholism. 23:3-18.

Baker PT. (Ed). 1978. The Biology of High Altitude Peoples. Cambridge University Press.

Baker JR, Callan HG. 1950. Heterochromatin. Nature. 166:227.

Baliček P, Zizka J, Skalska H. 1977. Length of human constitutive heterochromatin in relation to chromosomal contraction. Hum Genet. 38:189-193.

Baliček P, Zizka J, Skalska H. 1978. Variability and familial transmission of constitutive heterochromatin of human chromosomes evaluated by method of linear measurement. Hum Genet. 42:257-256.

Bardham S, Singh DN, Davis K. 1981. Polymorphism in chromosome 4. Clin Genet. 20:44-47.

Beall CM. 2000. Tibetan and Andean patterns of adaptation to high-altitude hypoxia. Hum Biol. 72(1):201-228.

Beall CM, Blangero Y, Williams-Blangero S, et al. 1994. A major gene for percent of oxygen saturation of arterial hemoglobin in Tibetan highlandrs. Am J Phys Anthropol. 95:271-276.

Beall CM, Cavalleri GL, Deng L, et al. 2010. Natural selection on EPAS1 (HIF2alpha) associated with low hemoglobin concentration in Tibetan highlanders. Proc Natl Acad Sci USA. 107:11459-11464.

Belmont AS, Zhai Y, Thilenius A. 1993. Lamin B distribution and association with peripheral chromatin revealed by optical sectioning and electron microscopy tomography. J. Cell Biol. 123:1671-1685.

Belmont AS, Bignone F, Ts'o POP. 1986. The relative intranuclear positions of Barr bodies in XXX non-transformed human fibroblasts. Exp Cell Res, 165(1):165-79.

Benezech M, Robert B, Noël B, Zintilini R. 1973. Agressivite, delinquence psychopatique et longeur du chromosome Y. Nouv Presse Med. 2:583.

Bengtsson B, Gustavson KH, Reuterskiöld G, Santesson B, Ahnsen S. 1974. Male pseudohermaphroditism with 45 X/46XYq-mosaicism in a pair of monozygotic twins. Clin Genet. 5:133.

Beridze TG. 1982. Satellite DNA. The science. Moscow (in Russian).

Bernard C. 1872. Physiologic Generale. Librairie Hachette. Paris. France.

Bernardi G, Bernardi G. 1986. Compositional constraints and genome evolution. J Mol Evol. 24:1-11.

Bertalanfy VL. 1968. General System Theory. Foundations, Development, Applications. New York.

Bhasin MK. 2005. Human population cytogenetics. A review. Int J Hum Genet. 5(2):83-152.

Bigham AW, Mao X, Mei R, et al. 2009. Identifying positive selection candidate loci for high-altitude adaptation in Andean populations. Hum Genomics. 4:79-90.

Bigham A, Bauchet M, Pinto D, et al. 2010. Identifying signatures of natural selection in Tibetan and Andean populations using dense genome scan data. PLoS Genet. 6:e1001116.

Bishop A, Blank CT, Hunter H. 1962. Heritable variation in the length of the Y chromosome. Lancet. ii:18.

Blatties CM. 1997. Thermoregulation: 10th International Symposium on the Pharmacology of Thermoregulation. Ann New York Acad Sci. Book 813.

Blumberg B, Shulkin JD, Rotter JI, et al. 1980. Habitual abortion and heterochromatic chromosomal polymorphisms. Am J Hum Genet. 32:63A.

Blumberg B, Shulkin JD, Rotter JI, et al. 1982. Minor chromosomal variants and major chromosomal anomalies in couples with recurrent abortion. Am J Hum Genet. 34:948.

Bobrow M, Pearson PL, Gollacott HEAC. 1971a. Para-nucleolar position of the human Y chromosome in interphase nuclei. Nature. 232:556-557.

Bobrow M, Pearson PL, Pike MC, El-Alfi OS. 1971b. Length variation in the quinacrine-binding segment of human Y chromosomes of different sizes. Cytogenetics. 10:190-198.

Bobrow M, Madam K, Pearson PL. 1972. Staining of some specific regions of human chromosomes, particularly the secondary constriction of No 9. Nature New Biol. 238(82):122-124.

Borisov YuM. 1980. Geographical variability of variants of the system of additional chromosomes in continental forms of *Apodemus peninsulae* (Rodentia, Muridae). Izv. SB USSR Academy of Sciences, ser. "Biology", vol. 3:61-69 (in Russian).

Bostock CJ, Sumner AT. 1978. The eukaryotic chromosomes. North-Holland Publ. Company. Amsterdam-New York-Oxford.

Bostock CJ, Gosden JR, Mitchell AR. 1978. Localization of a male-specific DNA fragment to human Y chromosomes. Nature. 272:324-328.

Bott CE, Sekhon GS, Lubs HA. 1975. Unexpected high frequency of paternal origin of trisomy 21. Am J Hum Genet. 24:20A.

Bourgeois CA, Laquerriere F, Hemon D, Hubert J, Bouteille M. 1985. New data on the in-situ position of the inactive X chromosome in the interphase nucleus of human fibroblasts. Hum Genet. 69(2):122-29.

Boveri Th. 1904. Ergebnisse über die constitution der chromatischen substanz des Zollkerns. Fischer, Jena, 1904.

Boyle S, Gilchrist S, Bridger JM, et al. 2001. The spatial organization of human chromosomes within the nuclei of normal and emerin-mutant cells. Hum Mol Genet. 10:211-219.

Bray GA, Bouchard C. 1997. Genetics of human obesity: research directions. FASEB J. 11:937-945.

Breg WR, Allderdice PW, Miller DA, Miller DJ. 1972. Quinacrine fluorescence patterns and terminal DNA labeling of human C group chromosomes. Nature New Biol. 236(64):76-77.

Britten RJ, Kohne DF. 1968. Repeated sequences in DNA. Science. 161:529-540.

Brogger A, Urdal T, Larsen FB, Lavik NJ. 1977. No evidence for a correlation between behavior and the size of the Y chromosome. Clin. Genet. 11:349.

Brown SW. 1966. Helerochromatin. Science. 131:417-425.

Buckton KE, O'Riordan ML, Jacobs PA, et al. 1976. C- and Q-band polymorphisms in the chromosomes of three human populations. Ann. Hum. Genet. 40:90-112.

Buzin CH, Mann JR, Singer-Sam J. 1994. Quantitative RT-PCR assays show *Xist* RNA levels are low in mouse female adult tissues, embryos and embryoid bodies. Development. 120:3529-3536.

Carnevale A, Ibanez BB, Castillo V. 1976. The segregation of C-band polymorphisms on chromosomes 1, 9 and 16. Am J Hum Genet. 28:412-416.

Caspersson T, Farber S, Foley GE, et al. 1968. Chemical differentiation along metaphase chromosomes. Exp Cell Res. 49(1):219-222.

Caspersson T, Zech L, Modest EJ, et al. 1969a. Chemical differentiation with fluorescent alkylating agents in Vicia faba metaphase chromosomes. Exp Cell Res. 58(1):128-140.

Caspersson T, Zech L, Modest EJ, et al. 1969b. DNA-binding fluorochromes for the study of the organization of the metaphase nucleus. Exp Cell Res. 58(1): 141-152

Caspersson T, Zech L, Johansson C. 1970. Differential binding of alkilating fluorochromes in human chromosomes. Exp Cell Res. 60:315-319.

Cervenka J, Jacobson DT, Gorlin RJ. 1971. Fluorescing structures of human metaphase chromosomes. Detection of "Y body". Amer J Hum Genet. 23:317.

Chagnon YC, Bouchard C. 1996. Genetics of obesity: advances from rodent studies. Trends Genet. 12:441-444.

Chaligne R, Heard E. 2014. X-chromosome inactivation in development and cancer. FEBS Lett. 588(15):2514-22.

Chandley AC, Edmond P. 1971. Meiotic studies on a subfertile patient with a ring Y chromosome. Cytogenetics. 10:295.

Chen T. 1977. Fluorescent Q bands of human chromosomes with Hoechst 33258 stain.
Humangenetik. 36:283-288.

Chiarelli B, Lin CC. 1972. Comparison of florescence patterns in human and chimpanzee chromosomes. Genet Phaenen 15:103-106.

Clement K, Vaisse C, Guy-Grand B, Froguel P. 1998. A mutation in the human leptin receptor gene causes obesity and pituitary dysfunction. Nature. 392:398-401.

Cohen MM, Shaw MW, Macluer JW. 1966. Racial differences in the human Y chromosome. Cytogenetics. 5:34-52.

Comings DE. 1968. The rational for an ordered arrangement of chromatin in the interphase nucleus. Amer J Hum Genet. 20:440.

Comings DE. 1971. Heterochromatin of the Indian muntjac. Exp Cell Res. 67:441-460.

Comings DE. 1972. The structure and function of chromatin. Adv Hum Genet. 3:237-431.

Comings DE. 1973. Biochemical mechanisms of chromosome banding and color banding with acridine orange. In: Chromosome identification—technique and application in biology and medicine. New York, London, pp. 293-299.

Comings DE, Okada TA. 1973. DNA replication and the nuclear membrane. J Mol Biol. 75:609-618.

Commoner B. 1964. Roles of deoxyribonucleic acid in inheritance. Nature. 202:960-968.

Conen PE, Baley JD, Allemang WH, et al. 1961. A probable partial deletion of the Y chromosome in intersex patient. Lancet. 2:294.

Corneo G, Ginelly E, Polli E. 1967. A satellite DNA isolated from human tissues. J Mol Biol. 23:619-622.

Corneo G, Ginelli E, Polli E. 1968. Renaturation properties and localization in heterochromatin of human satellite DNAs. Biochem Biophys Acta. 247:528-534.

Court Brown WM. 1967. Human population cytogenetics. Amsterdam: North-Holland.

Court Brown WM, Jacobs PA, Brunton M. 1965. Chromosome studies on randomly chosen men and women. Lancet. ii: 561-562.

Craig-Holmes AP, Shaw MW. 1971. Polymorphism of human constitutive heterochromatin. Science. 174:702-704.

Craig-Holmes AP, Moore F, Shaw M. 1975. Polymorphism of human C-band heterochromatin. II. Family studies with suggestive evidence for somatic crossing over. Am J Hum Genet. 27:178-189.

Cremer T, Cremer C. 2001. Chromosome territories, nuclear architecture and gene regulation in mammalian cells. Nat Rev Genet. 2:292-301.

Cremer T, Cremer C, Bauman H. 1982. Rabl's model of the interphase chromosome arrangement tested in Chinese hamster's cell by premature chromosome condensation and laser-UV-microbeam experiments. Hum Genet. 60:45-56.

Cremer K, Kreth G, Koester H, et al. 2000. Chromosome territories, interchromatin domain compartment, and nuclear matrix: an integrated view of functional nuclear archtecture. Crit Rev Eukaryot Gene Expr. 10:179-212.

Croft JA, Bridger JM, Boyle S, et al. 1999. Differences in the localization and morphology of chromosomes in the human nucleus. J Cell Biol. 145:1119-1131.

Dahlgren L. 1978. Female alcoholic. Acta Psychiat Scand. 57:325-335.

Darnell JE. Jr. 1978. Implications of RNA-RNA splicing in evolution of eukaryotic cells. Science. 202:1257-1260.

Darlington C.D. 1937. Recent advances in cytology. 2nd ed. Blakiston.

Darlington CD, La Cour L. 1940. Nucleic acid starvation of chromosomes in Trillium. J Genet. 40:185-213.

de 1a Chapelle A. 1961. Constrictions in normal human chromosomes. Lancet. ii: 460-463.

de 1a Chapelle AP, Aula P, Kivalo E. 1963. Enlarged short arm or satellite region—a heritable trait probably unassociated with developmental disorder. Cytogenetics. 2:129-139.

de la Chapelle A, Schroder J, Stenstrand K. et al. 1974. Pericentric inversions of human chromosomes 9 and 10. Am J Hum Genet. 26:746-766.

Décsey K, Bellovits O, Bujdoso GM. 2006. Human chromosomal polymorphism in Hungarian sample. Int J Hum Genet. 6(3):177-183.

Denver Conference. 1960. A proposed standard system of nomenclature of human mitotic chromosomes. Lancet. i: 1063-1065.

Dev VG, Warburton D, Miller OJ, et al. 1972. Consistent pattern of binding of antiadenosine antibodies to human metaphase chromosomes. Exp Cell Res. 74:288-293.

Dietzel S, Welland E, Ellis R, Munkel C, Cremer C, Cremer T. 1995. Three dimensional distribution of centromeric or paracentromeric heterochromatin of chromosomes 1, 7, 15 and 17 in human lymphocyte nuclei studied with light microscopic axial tomography. Bioimaging. 3:121-133.

Distèche C, Bontemps J. 1974. Chromosome regions containing DNAs of known base composition specifically evidenced by 2,7-dibutyl proflavine. Chromosoma. 47:263-281.

Doolitle WF, Sapienza C. 1980. Selfish genes, the phenotype paradigm and genome evolution. Nature. 284:601-603.

Dragoni TA, Zeng ZB, Canzian F, et al. 1995. Mapping of body weight loci on mouse chromosome X. Mamm Genom. 6:778-781.

Dyer AF. 1964. Heterochromatin in American and Japanese species of Trillium. II. The behavior of H segments. Cytologia. 29:171-190.

Eiberg H. 1974. New selective giemsa technique for human chromosomes, Cd staining. Nature. 248:55.

Eils RC, Dietzel S, Bertin E. et al. 1996. Three-dimensional reconstruction of painted human interphase chromosomes: active and inactive X chromosome territories have similar volumes but differ in shape and surface structure. J Cell Biol. 135:1427-1440.

Ellenberg J. Siggia ED, Mereira JE, et al. 1997. Nuclear membrane dynamics and reassembly in living cells: targeting of an inner nuclear membrane protein in interphase and mitosis. J Cell Biol. 138:1193-1206.

Ellison JR, Barr HJ. 1972. Quinacrine fluorescence of specific chromosome regions. Late replication and high A:T content in *Samoaia leonensis*. Chromosoma. 36:375.

Erdtmann B. 1982. Aspects of evaluation, significance, and evolution of human C-band heteromorphism. Hum Genet. 61:281-294.

Erdtmann B, Salzano F. Mattevi MS. 1987. Size variability of the Y chromosome distal C- band in Brazilian Indians and Caucasoids. Ann Hum Biol. 8:415-424.

Erikson CK. 1979. Factors affecting the distribution and measurement of ethanol in the body. In: Majchrowicz, E. and Noble, E.P. (eds.). "Biochemistry and phamaco1ogy of ethanol". Plenum Press, New York, London. 1:9-23.

Erickson YC, Hollopeter G, Palmiter RD. 1996. Attenuation of the obesity syndrome of ob/ob mice by the loss of neuropeptide Y. Science. 274:1704-1707.

Evans HJ. 1973. Molecular architecture of human chromosomes. Brit Med Bull. 29:196-202.

Evans HJ, Buckton KE, Sumner AT. 1971. Cytological mapping of human chromosomes: results obtained with quinacrine fluorescence and the acetic-saline-giemsa techniques. Chromosoma. 35:310.

Evans HJ, Buckland RA, Pardue ML. 1974. Location of the genes coding for 18 S and 28 S ribosomal RNA in the human genome. Chromosoma. 48:405-426.

Fan W, Boston BA, Kesterson RA, Hruby KУ, Cone RD. 1997. Role of melanocortinergic neurons in feeding and the agouti obesity syndrome. Nature. 385:165-168.

Ferguson-Smith MA, Ferguson-Smith ME, Ellis PM, Dickson M. 1962. The sites and relative frequencies of secondary constrictions in human somatic chromosomes. Cytogenetics. 1:325-343.

Fergusson-Smith MA, Boyd E, Fergusson-Smith ME. et al. 1969. Isochromosome for long arm of Y chromosome in patient with Turner's syndrome and sex chromosome mosaicism (45,X/46, XYqi). J Med Genet. 6:422.

Ferreira J, Paolella G, Ramos C, Lamond AI. 1997. Spatial organization of large-scale chromatin domains in the nucleus: a magnified view of single chromosome territories. J Cell Biol. 139:1597-1610.

Ferrell RE, Bertin T, Young R. et al. 1978. The Aymara of Western Bolivia. IV. Gene frequencies of eight blood groups and 19 protein and erythrocyte enzyme systems. Am J Hum Genet. 30:539-549.

Flamm WG, Walker MB, McCallum M. 1969. Some properties of the single strands isolated from the DNA of the nuclear satellite of the mouse (*Mus musculus*). J Mol Biol. 40:423-443.

Fogle TA, McKenzie WH. 1980. Cytogenetic study of a large black kindred: inversions, heteromorphisms and segregation analysis. Hum Genet. 55:345.

Folk GE. 1974. Textbook of Environmental Physiology. Lea & Febiger. Philadelphia.

Ford CE, Hamerton JH. 1956. The chromosomes of man. Nature, 178:1030-1023.

Ford CE, Jones KW, Polani PE. 1959. A sex-chromosome anomaly in a case of gonadal disgenesis (Turner's syndrome). Lancet. i: 711-713.

Franke WW, Krien S. 1972. Nuclear membrane attached DNA enriched in repetitive sequences. Naturwissenschaften 59:37.

Fredga K, Mandahl N. 1972. Autosomal heterochromatin in some carnivores. In: Chromosome Identification. Nobel Symp. No. 23. T. Caspersson and L. Zech. (eds.). London: Academic Press. pp. 104-167.

Friedman JM. 2000. Obesity in the new millennium. Nature. 404:632-634.

Friedman JM, Halaas JL. 1998. Leptin and the regulation of body weight in mammals. Nature. 395:763-770.

Fussel CP. 1975. The position of interphase chromosomes and late replicating DNA in centromere and telomere regions of *Allium cepa L.* Chromosoma. 50:201-210.

Gagne R, Laberge C. 1972. Specific cytological recognition of the heterochromatic segment of number 9 chromosome in man. Exp Cell Res. 73(1):239-242.

Gagne R, Laberge G, Tanguay R. 1972. Interphase association of human "Y body" with nucleolus. Johns Hopk Med J. 130:254-258.

Gall JG, Pardue ML. 1969. Formation and detection of RNA-DNA hybrid molecules in cytological preparations. Proc Natl Acad Sci. USA, 63(2):378-383.

Galli J, Li LS, Glaser A, et al. 1996. Genetic analysis of non-insulin dependent mellitus in the GK rat. Nature Genet. 12:31-37.

Galupa R, Heard E. 2018. X-chromosome inactivation: a crossroads between chromosome architecture and gene regulation. Annu Rev Genet. 52:535-66.

Ganner E, Evans HJ. 1971. The relationship between patterns of DNA replication and of quinacrine fluorescence in the human chromosome complement. Chromosoma. 35(3):326-341.

Gauguier D, Froguel P, Parent V. et al. 1996. Chromosomal mapping of genetic loci associated with non-insulin dependent diabetes in the GK rat. Nature Genet. 12:38-43.

Genest P. 1979. Chromosome variants and abnormalities detected in 51 married couples with repeated spontaneus abortions. Clin Genet. 16:387-389.

Genest P, Genest FB. 1985. The influence of the length of the human Y-chromosome on spontaneous abortions. A perspective study in family lines with inherited polymorphic Y-chromosomes. Ann Genet. 28(3):143-148.

Gerace L, Burke B. 1988. Functional organization of the nuclear envelope. Annu Rev Cell Biol. 4:335-374.

Geraedts JPM, Pearson P L. 1974. Fluorescent chromosome polymorphism: frequencies and segregation in a Dutch population. Clin Genet. 6:247-257.

German J. 1964. The pattern of DNA synthesis in the chromosomes of human blood cells. J Cell Biol. 20:37-55.

Gershenson SM. 1933. Studies on the genetically inert region of the X-chromosome of Drosophila. I. Behavior of an X-chromosome deficient for a part of its inert region. J Genet. 28:297-313.

Ghosh PK, Singh IP. 1975. Morphological variability of the human chromosomes in two Indian populations-Rajputs and Punjabis. Humangenetik. 29:67.

Ginsburg EKh, Axenovich TI, Babenko VN. 1991. Comparison of asymptotic tests of the Hardy-Weinberg distribution. Ann Hum Genet. 55(4):329-343.

Goodpasture C, Bloom SE. 1975. Visualization of nucleolar organizer regions in mammalian chromosomes using silver staining. Chromosoma 53:37-50.

Goodwin DIW. 1988. Is alcoholism hereditary? Second Edition. Ballantine Books; New York.

Gosden JR, Mitchell AR, Buckland RA, et al. 1975. The location of four human satellite DNAs on human chromosomes. Exp Cell Res. 92:148-158.

Gottesfeld JM, Bonner J, Radda GK, Walker IO. 1974. Biophysical studies on the mechanism of quinacrine staining of chromosomes. Biochemistry. 13:2937-2945.

Gripenberg U. 1965. Size variation and orientation of the human Y chromosome. Chromosoma. 15:618.

Gruzdev AD. 2000. DNA topology in heterochromatin (a hypotheses). J Theor Biol. 207:255-264.

Halbrecht I, Shabtai F. 1976. Human chromosome polymorphisms and congenital malformations. Clin Genet. 10:113-122.

Hancock AM, Witonsky DB, Alkorta Aranburu G, et al. 2011. AdaptationstoClimate-MediatedSelectivePressuresinHumans.PLoSGenet. 7(4):e1001375.

Harrison GA, Weiner JS, Tanner JM, Barnicot NA. 1977. Human Biology. Oxford, Oxford University Press.

Harrison GA. 1997. The role of the Human Adaptability International Programme in the development of human population biology. In: Human Adaptability: Past, Present and Future. Ulijaszek T.M and Huss-Ashmore R.A. (eds.). Oxford, England: Oxford University Press. 17-25.

Hatch FT, Bodner AJ, Mazrimad JA, Moor DH. 1976. Satellite DNA and cytogenetic evolution, DNA quantity, satellite DNA karyotypic variations in Kangaroo rats (Genus Dipodomys). Chromosoma. 58:155-168.

Heitz E. 1928. Das Heterochromatin der Moose. J Jahrb Wisenschs Bot. 69:762-818.

Heitz E. 1934. Die Somatische heteropyknose bei Drosophila melanogaster und ihre genetische Bedeutung. Z Zellfosch. 20:237-287.

Heitz E. 1935. Chromosomen structur und Gene. Ztschr. induct. Abstammungs und vererbungslehre 70:402-447.

Hemming L, Burns C. 1979. Heterochromatic polymorphism in spontaneus abortions. J Med Genet. 16:358-362.

Henderson AS, Warburton D, Atwood KC. 1972. Location of ribosomal DNA in the human chromosome complement. Proc Nat Acad Sci USA. 69:3394-3398.

Herva R. 1981. Q and C band chromosomal heteromorphisms in Northern Finnish population. Doctoral thesis. Univ. of Oulu: Oulu.

Hoehn H, Martin GM. 1973. Nonrandom arrangement of human chromatin: topography of disomic markers X, Y, and 1h+. Cytogenet Genome Res. 12(6):443-52.

Holmquist GP, Dancis B. 1979. Telomere replication kinetochore organizers and satellite DNA evolution. Proc Nat Acad Sci USA. 76:4566-4570.

Hotta Y, Stern H. 1978. Absence of satellite DNA synthesis during meiotic prophase in mouse and human spermatocytes. Chromosoma. 69:323.

Howell WM, Denton FE, Diamont TR. 1975. Differential staining of the satellite regions of human acrocentric chromosomes. Experientia. 31:260-262.

Hsu TC. 1975. A possible function of constitutive heterochromatin: the bodyguard hypothesis. Genetics. 79:137-150.

Hsu TC. 1979. Human and mammalian cytogenetics. An historical perspective. Springer, Heidelberg.

Hsu TC, Arrighi FE. 1971. Distribution of constitutive heterochromatin in mammalian chromosomes. Chromosoma, 34:243-253.

Ibraimov AI, Mirrakhimov MM, Nazarenko SA, Axenrod EI, Akbanova GA. 1982. Human chromosomal polymorphism. I. Chromosomal Q-polymorphism in Mongoloid populations of Central Asia. Hum Genet. 60:1-7.

Ibraimov AI, Mirrakhimov MM. 1982a. Human chromosomal polymorphism. II. Chromosomal C-polymorphism in Mongoloid populations of Central Asia. Hum Genet. 60:8-9.

Ibraimov AI, Mirrakhimov MM. 1982b. Human chromosomal polymorphism. III. Chromosomal Q-polymorphism in Mongoloids of Northern Asia. Hum Genet. 62:252-257.

Ibraimov AI, Mirrakhimov MM. 1982c. Human chromosomal polymorphism. IV. Q-polymorphism in Russians living in Kirghizia. Hum Genet. 62:258-260.

Ibraimov AI, Mirrakhimov MM. 1982d. Human chromosomal polymorphism. V. Chromosomal Q-polymorphism in African populations. Hum Genet. 62:261-265.

Ibraimov AI. 1983. Chromosome preparations of human whole lymphocytes - an improved technique. Clin Genet. 24:240-242.

Ibraimov AI, Mirrakhimov M M. 1985. Q-band polymorphism in the autosomes and the Y chromosome in human populations. In: "Progress and Topics in Cytogenetics. The Y chromosome. Part A. Basic characteristics of Y chromosome". A. A. Sandberg (Ed). Alan R. Liss, Inc., New York. USA. 213-287.

Ibraimov AI, Mirrakhimov MM, Axenrod EI, Kurmanova G.U. 1986. Human chromosomal polymorphism. IX. Further data on the possible selective value of chromosomal Q-heterochromatin material. Hum Genet. 73:151-156.

Ibraimov AI, Kurmanova GU, Ginsburg EK, Aksenovich TI, Axenrod EI. 1990. Chromosomal Q-heterochromatin regions in native highlanders of Pamir and Tien-Shan and in newcomers. Cytobios. 63:71-82.

Ibraimov AI, Axenrod EI, Kurmanova GU, Turapov OA. 1991. Chromosomal Q-heterochromatin regions in the indigenous population of the Northern part of West Siberia and in new migrants. Cytobios. 67:95-100.

Ibraimov AI. 1993. The origin of modern humans: a cytogenetic model. Hum Evol. 8(2):81-91.

Ibraimov AI, Karagulova GO, Kim EY. 1997. Chromosomal Q-heterochromatin regions in indigenous populations of the Northern India. Ind. J. Hum. Genet. 3: 77-81.

Ibraimov AI, Karagulova GO, Kim EY. 2000. The relationship between the Y chromosome size and the amount of autosomal Q-heterochromatin in human populations. Cytobios. 102:35-53.

Ibraimov AI. 2003. Condensed chromatin and cell thermoregulation. Complexus. 1:164-170.

Ibraimov AI. 2004. The origin of condensed chromatin, cell thermoregulation and multicellularity. Complexus. 2:23-34.

Ibraimov AI, Karagulova GO. 2006a. Chromosomal Q-heterochromatin regions in individuals of various age groups. Int J Hum Genet. 6(3):219-228.

Ibraimov AI, Karagulova GO. 2006b. Chromosomal Q-heterochromatin variability in neonates deceased during first year of age. Int J Hum Genet. 6(4):281-285.

Ibraimov AI, Tabaldiev SK. 2007. Condensed chromatin, cell thermoregulation and human body heat conductivity. J Hum Ecol. 21(1):1-22.

Ibraimov AI. 2007. The evolution of body heat conductivity, skin and brain size in human. J Hum Ecol. 21(2):95-103

Ibraimov AI. 2008. On the pathogenesis of atherosclerosis: the role of the blood temperature, phase transitions of lipoproteins and turbulence of blood flow. J Hum Ecol. 23(3):195-201.

Ibraimov AI. 2010. Chromosomal Q-heterochromatin regions in populations and human adaptation. In: MK Bhasin, C Susanne (Eds.): Anthropology Today: Trends and Scope of Human Biology. Delhi: Kamla-Raj Enterprises, pp. 225-250.

Ibraimov AI, Kazakova AK, Moldotashev IK, Sultanmuratov MT, Abdyev KS. 2010a. Variability of Human Body Heat Conductivity in Population. I. Methodological and Theoretical Approaches. J Hum Ecol 32(1):1-22.

Ibraimov AI, Kazakova AK, Moldotashev IK, Sultanmuratov MT, Abdyev KS. 2010b. Variability of Human Body Heat Conductivity in Population. II. Diseases of Civilization. J Hum Ecol. 32(2):69-78.

Ibraimov AI. Origin of modern humans: a cytogenetic model. Hum Evol. 2011a, 26(1-2):33-47.

Ibraimov AI. 2011b. Evolution without genes. Int J Genet. 3(1):50-61.

Ibraimov AI. 2012. On the origin of sex. GJSFR (C), 12(8) (Ver. 1.0):1-6. [Global Journal of Science Frontier Research: C Biological Sciences].

Ibraimov AI, Akanov AA, Meymanaliev TS, Karakushukova AS, Kudrina NO, Sharipov KO, Smailova RD. 2013. Chromosomal Q-heterochromatin polymorphisms in 3 ethnic groups (Kazakhs, Russians and Uygurs) of Kazakhstan. Int J Genet. 5(1):121-124.

Ibraimov AI. 2014. Chromosomal Q-heterochromatin and sex in human population. J Mol Biol Res. 4(1):10-19.

Ibraimov AI, Akanov AA, Meymanaliev TS, Smailova RD, Baygazieva GD. 2014a. Chromosomal Q-heterochromatin and age in human population. J Mol Biol Res. 4(1):1-9.

Ibraimov AI, Akanov AA, Meimanaliev TS, Sharipov KO, Smailova RD, Dosymbekova R. 2014b. Human Chromosomal Q-heterochromatin Polymorphism and Its Relation to Body Heat Conductivity. Int J Genet. 6(1):142-148.

Ibraimov AI. 2015a. Heterochromatin: The visible with many invisible effects. Global Journal of Medical Research (C), 15(3):7-32.

Ibraimov AI. 2015b. The Evolution of Material Basis of Evolution. J Adv Biol. 8(2):1596-1607.

Ibraimov AI. 2016a. Methodical Approaches to the Study of Human chromosomal Q-Heterochromatin Variability. J Mol Biol Res. 6(1):1-10.

Ibraimov AI. 2016b. Chromosomal Q-Heterochromatin Polymorphism in Patients with Alimentary Obesity. Biol. Med. (Aligarh). 8: 275. DOI: 10.4172/0974-8369.1000275

Ibraimov AI. 2016c. Chromosomal Q-heterochromatin Regions in Alcoholics and Drug Addicts. Biol. Med. (Aligarh). 8:346. DOI: 10.4172/0974-8369.1000346

Ibraimov AI. 2017a. Cell Thermoregulation: Problems, Advances and Perspectives. J Mol Biol Res. 7(1):58-79. doi:10.5539/jmbr.v7n1p58

Ibraimov AI. 2017b. From 48 to 46 chromosomes: Origin of Man. J Mol Biol Res. 7(l):80-87. doi:10.5539/jmbr.v7n1p80

Ibraimov AI. 2017c. Chromosomal Q-Heterochromatin and Atherosclerosis. J Mol Biol Res. 7(1):1. doi:10.5539/jmbr.v7n1p143

Ibraimov AI. 2018a. Why Are All People Not Ill with High-Altitude Pulmonary Edema? J Cardiol Diagnost Res. 1(1):13-18.

Ibraimov AI. 2018b. Human Body Heat Conductivity in norm and pathology: A review. Advance Research Journal of Multidisciplinary Discoveries. 32(3):12-21.

Ibraimov AI. 2019a. Human adaptation: why only genes? Int J Biol Med. 1:22-33.

Ibraimov AI. 2019b. The origin of modern humans. What was primary: genes or heterochromatin? Hum Evol. 34(1-2):39-58.

Ibraimov AI. 2019c. Chromosomal Q-heterochromatin Polymorphism and Sport. Int J Biol Med. 1:45-52.

Ibraimov AI. 2019d. X-chromosome inactivation: dosage compensation of genes or heterochromatin? Int J Biol Med. 1:75-87.

Ibraimov AI. 2019e. Cell Thermoregulation: Reality or just a logical Construction? Int J Biol Med. 1:65-74.

Ibraimov A.I. 2019f. Cell thermoregulation hypothesis: its origin, material basis, mechanisms and meaning. Int J Biol Med. 1:114-123.

Ibraimov A.I. 2019g. Cell thermoregulation and origin of homoeothermic animals, Current Research in Biochemistry and Molecular Biology, 1(1) 10-13.

Ibraimov A.I. 2019h. Evolution Without Genes: A Review, Current Research in Biochemistry and Molecular Biology, 1(1),45-68.

Ibraimov A.I. 2020a. The origin of the human karyotype: its uniqueness, causes and effects, Current Research in Biochemistry and Molecular Biology, 1, 1, 9-20.

Ibraimov A.I. 2020b. Cell thermoregulation: How is excess heat eliminated?, Current Research in Cytology and Histology 1(1), 14-21.

Ibraimov A.I. 2020c. Origin of Species by Means of Adaptation, or the Struggle for Life by "Active" and "Passive Darwinism". Canadian Journal of Biomedical Research and Technology, 3(2): 1-6.

Inuma K, Matsunaga E, Nakagome J. 1973. Polymorphism of Q ana C bands in human chromosomes. Annu Rep Natl Genet (Japan) 23:112.

ISCN. 1978. An international system for human cytogenetic nomenclature. Report of the standing committee on human cytogenetic nomenclature. Cytogenet Cell Genet. 21:313-404.

ISCN. 2009. An international system for human cytogenetic nomenclature. Eds.: L.G. Shaffer, M.L. Slovak, L.J. Campbell. Karger.

Jacobs PA. 1977. Human chromosome heteromorphisms (variants). In: Progress in Medical Genetics, Vol. II. Steinberg A.G., Beam A.G. and Motulsky A.G. (eds.). Philadelphia. Saunders. 251-272.

Jacobs PA, Ross A. 1966. Structural abnormalities of the Y chromosome in man. Nature. 210: 352.

Jacobs PA, Baikie AG, Court Brown WM. 1959. Evidence for the existence of the human “super female”. Lancet. ii: 423-425.

Jacobs PA, Cruickshank MJW, Faed A, et al. 1967. Pericentric inversion of a group C autosome; a study of three families. Ann Hum Genet. 31:219-230.

Jacobs PA, Frackiewicz A, Law P, et al. 1975. The effect of structural aberrations of the chromosomes on reproductive fitness in man. II. Results. Clin Genet. 8:169-178.

James W. 1996. In: The Origins and Consequences of obesity (eds. Chadwick D. and Cardew G.). 1-16, Wiley, Chichester.

Jeong C, Witonsky DB, Basnyat B, et al. 2018. Detecting past and ongoing natural selection among ethnically Tibetan women at highaltitude in Nepal. PLoSGenet. 14:9.

John B. 1988. The biology of heterochromatin. In: “Heterochromatin: Molecular and Structural Aspects”. Verma R.S. (Ed.). Cambridge University Press, Cambridge, New York, New Rochelle, Melburn, Sydney. pp.1-147.

John B, Micloš GLG. 1979. Functional aspects of satellite DNA and heterochromatin. Int Rev Cytol. 58:1-114.

John HA, Birnstiel ML, Jones KW. 1969. RNA-DNA hybrids at the cytological level. Nature. 223(5206):582-587.

Jones KW. 1970. Chromosomal and nuclear location of mouse satellite DNA in individual cells. Nature. 225:215-215.

Jones RW, Corneo G. 1971. Location of satellite and homogenous DNA sequences on human chromosomes. Nature New Biol. 233(43):268-271.

Jones KW, Purdon IF, Prosser J, Corneo G. 1974. The chromosome localization of human satellite DNA I. Chromosoma. 49:161-172.

Kalz L, Kalz-Fuller B, Hedge S, Schwanitz G. 2005. Polymorphism of Q-band heterochromatin; qualitative and quantitative analyses of features in 3 ethnic groups (Europeans, Indians, and Turks). Int J Hum Genet. 5(2):153-163.

Käosaar ME, Mikelsaar A.-V. 1973. Chromosome investigation in married couples with repeated spontaneous abortions. Humangenetik. 17:277-283.

Keller M. 1970. The great Jewish mystery. British J Addict. 64:287-296.

Kendall MG, Stewart A. 1973. The advanced theory of statistics. Vol. 2. Interference and relationship. Charles Griffin and Company, London.

Khesin RB, Leibovich BA Chromosome structures, histones and gene activity in Drosophila. Mol Biol. 10:3-33 (in Russian).

Kikuchi Y, Sandberg A. 1965. Chronology and pattern of human chromosome replication. II. Autoradiographic behavior of various Y and X chromosomes. J Nat Cancer Inst. 34:795.

King MC, Wilson AC. 1975. Evolution at two levels in humans and chimpanzees. Science. 108:107-116.

Kivi SYu, Mikelsaar A.-V.N. 1981. Variants of Q- and C-segments of chromosomes in patients suffering from breast and ovarian cancer. In the book: Polymorphism of chromosomes in humans. AMS USSR, Moscow, pph. 196-204.

Klinger HP. 1958. The fine structure of the sex chromatin body. Exp Cell Res. 14(1):207-11.

Krmpotic E, Szego K, Modestas R, Molabola GB. 1972. Localization of male determining factor on short arm of Y chromosome. Case report of a baby with 46, X, t(Yp+; 14q-). Clin Genet. 3:381.

Kunkel LM, Smith KD, Boyer SH. 1976. Human Y-chromosome specific reiterated DNA. Science. 191:1189-1190.

Kunkel LM, Smith KD, Boyer SH. 1977. Analysis of human chromosome specific DNA in chromosome variants. Proc Nat Acad Sci USA. 74:1245-1249.

Kurnik NB, Radeliffe IE. 1962. Reaction between DNA and quinacrine and other antimalarias. J Lab Clin Med. 60:669.

Kruminia AR, Kroshkina VG, Reshetnikov AI. 1987. Polymorphism of Q-segments of chromosomes in couples with impaired reproductive function. Genetics. 23(3):544-547 (in Russian).

Kurmanova GU. 1991. Variability of Q-heterochromatic regions of human chromosomes in connection with the problem of adaptation to high mountains. Author. Cand. dissert., Siberian Branch of the USSR Academy of Sciences, Novosibirsk (in Russian).

Laberge C, Gagne R. 1971. Quinacrine mustard staining solves the lenth variations of the human Y chromosome. Johns Hopkins Med J. 128:79.

Langmaid H, Laurence KM. 1974. Deletion of the long arms of the Y chromosome with normal male development and intelligence. J Med Genet. 11:208.

Ledley RS, Lubs HA, Ruddle FH. 1972. Introduction to chromosome analysis. Comput Biol Med. 2:107.

Lelikova GP, Tsvetkova TG. 1976. On the question of polymorphism of aboutcentromeric heterochromatin in populations of normal people. Genetics. 12:136-141 (in Russian).

Lejeune J, Gautier M, Turpin R. 1959. Etude des chromosomes somatiques de neuf enfants Mongolians. Compt. Rend. 248:1721-1722.

Lewontin RC. 1974. The genetic basis of evolutionary change. New York and London. Columbia Univ. Press.

Lichter P, Cremer T, Borden J, Manuelidis L, Ward D.C. 1988. Delineation of individual human chromosomes in metaphase and interphase cells by in situ suppression hybridization using recombination DNA libraries. Hum Genet. 80:224-234.

Lima-de-Faria A. 1959. Differential uptake of tritiated thymidine into heterochromatin and euchromatin in Melanoplus and Secale. J Biophys Biochem Cyt. 6:457-466.

Lima-de-Faria A. 1983. Molecular Evolution and Organization of the Chromosomes. Elsevier, Amsterdam, New York, Oxford.

Lima-de-Faria A. 1988. Evolution without Selection. Form and function by autoevolution. Elsevier. Amsterdam, New York, Oxford.

Lin CC, Gedeon MM, Griffith MM, et al. 1976. Chromosome analysis on 930 concecutive newborn children using quinacrine fluorescent banding technique. Hum Genet. 31:315-328.

Lin CC, Uchida IA, Byrnes EA. 1971. A suggestion for the nomenclature of the fluorescent banding patterns in human metaphase chromosomes. Can J Genet Cytol. 13:361.

Little MA, Garruto RM. 2000. Human adaptability research into the beginning of the third millennium. Hum Biol. 72 (1):179-199.

Lubs HA, Ruddle FH. 1971. Chromosome polymorphism in American Negro and White populations. Nature. 233:134-136.

Lubs HA, Patil S. 1975. Mediterranean origin of long Y chromosomes in Caucasians. Am J Hum Genet. 27:60 A.

Lubs HA, Patil SR, Kimberling WJ, et al. 1977. Racial differences in the frequency of Q- and C-chromosomal heteromorphism. Nature. 268:631-632.

Luciani JM, Morazzani M-R, Stahl A. 1975. Pachytene mapping of the C9 and acrocentric bivalents in the human oocyte. Chromosoma. 52:275-282.

Lyon MF. 1961. Gene action in the X-chromosome of mouse. Nature. 190:372-374.

Lyon MF. 1992. Some milestones in the history of X chromosome inactivation. Annu Rev Genet. 26:17-28.

Ma H, Seagal AJ, Berezney R. 1999. Association of chromosome territories with the nuclear matrix: distribution of human chromosome territories correlates with the release of a subset of nuclear matrix proteins. J Cell Biol. 146:342-352.

Madan K, Bobrow M. 1974. Structural variation in chromosome No 9. Ann Genet. 17:81-86.

Manuelidis L. 1978. Chromosomal localization of complex and simple repeated human DNAs. Chromosoma. 66:23-32.

Manuelidis L. 1982. Repeated DNA sequences and nuclear structures. In: Genome Evolution. Dover, G.A. and Flavell, R.B. (eds.). London: Academic Press. pp. 263-283.

Marahrens Y, Panning B, Dausman J, Strauss W, Jaenisch R. 1997. Xist-deficient mice are defective in dosage compensation but not spermatogenesis. Genes Dev. 11(2):156-66.

Marks J, Schmid EW, Sarich VM. 1988. DNA Hybridization as a Guide to Phylogeny
Relations of the Hominoidea. J Mol Evol. 17: 769-786.

Marshall WF, Dernburg AF, Harmon DA, Agard DA, Sedat JW. 1996. Specific interactions of chromatin with the nuclear envelope: positional determination within the nucleus in *Drosophila melanogaster*. Mol Biol Cell. 7:825-842.

Mather K. 1944. The genetic activity of heterochromatin. Proc Roy Soc London B. 132:308-332.

Mathew L. 1992. Severe Cold at High Altitude and Methods of Prevention of its III Effects. In: SK. Machanda, W. Selvamurthy, V. Mohan Kumar (eds.): Advances in Physiological Sciences. New Dehli, Macmillan India LTD, pp. 338-345.

Matsuura JS, Mayer M, Jacobs PA. 1979. A cytogenetic survey of an institution for the mentally retarded. III. Q-band chromosome heteromorphisms. Hum Genet. 52:203-210.

Mayr E. 1970. Populations, Species, and Evolution. The Belknap Press of Harvard
University Press, Cambridge, Massachusets.

Mayr E. 2000. Darwin's influence of modern thought. Sci Amer. July: 67-71.

Mayr E. 2002. What evolution is? Phoenix

Mazrimas JA, Hatch FT. 1972. A possible relationship between satellite DNA and evolution of kangaroo rat species (genus Dipodomys). Nature New Biol. 240:102-105.

McCracken AA, Daly PA, Zolnick MR, Clark AM. 1978. Twins and Q-banded chromosome polymorphisms. Hum Genet. 45:253-258.

McIlree ME, Price WH, Court Brown WM, et al. 1966. Chromosome studies on testicular cells from 50 subfertile men. Lancet. ii: 69.

McKay RDG, Bobrow M, Cooke HJ. 1978. The identification of a repeated DNA sequence involved in the karyotype polymorphism of the human Y chromosome. Cytogenet Cell Genet. 21:19-32.

McKenzie WH, Lubs HA. 1975. Human Q and C chromosomal variations: distribution and incidence. Cytogenet Cell Genet. 14:97-115.

McKenzie WH, Hosteller TL, Lubs HA. 1972. A family study: heritable variation in the length of the human Y chromosome. Am J Hum Genet. 24:616-693.

McKusick VA. 1980. The anatomy of human genome. J Hered. 71:370.

McKusick VA, Ruddle FH. 1977. The status of the gene map of the human chromosome. Science. 196:390-405.

Metaxotou C, Kalpini-Mavrou A, Panagae M, Tsenghi C. 1978. Polymorphism of chromosome 9 in 600 Greek subjects. Amer J Hum Genet. 30:85-89.

Meyer GF, Hess O, Beerman WS. 1961. Phasenspezifische Functionsstructuren in spermatocytenkernen von Drosophila melanogaster und ichre Abhängigkeit von Y-Chromosomen. Chromosoma. 12:676-716.

Micloš GLG. 1982. Sequencing and manipulating highly repeated DNA. In: Genome Evolution. G.A. Dover and R.B. Flavell (eds.). London: Academic Press, pp. 41-68.

Micloš GLG, John B. 1979. Heterochromatin and satellite DNA in man: properties and prospects. Am J Hum Genet. 31:264-280.

Mikelsaar AVN, Ilus T, Kivi S. 1978. Variant chromosome 3 (inv. 3) in normal newborns and parents and in children with mental retardation. Hum Genet. 41:109-113.

Mikelsaar AVN, Kääosaar ME, Tüür SJ, et al. 1975. Human karyotype polymorphism. III. Routine and fluorescence microscopic investigation of chromosomes in normal adults and mentally retarded children. Humangenetik 26:1-23.

Mikelsaar AVN, Schmidts M, Krone W, Schwarzacher HG, Schnedl W. 1977. Frequency of Ag-stained nucleolus organizer regions in the acrocentric chromosomes of man. Hum Genet. 37:73-77.

Mikelsaar AVN, Viikmao MN, Tuur SJ, Kaosaar ME. 1974. Human karyotype polymorphism. II. The distribution of individuals according to the presence of brilliant bands in chromosomes 3, 4 and 13 in a normal adult population. Humangenetik 23:59-63.

Mikelsaar AVN. 1979. Structural polymorphism of human chromosomes. Author. Dr. dissert., IMG AMN of the USSR, Moscow (in Russian).

Mikelsaar AVN. 1981. Study of human fluorescent chromosome variants. In the book: Polymorphism of chromosomes in humans. Medicine, Moscow (in Russian).

Miller D.A., Firschein 1. L., Dev V, G., Tantravahi R. and Miller O. J. 1974. The gorilla karyotype: chromosome length and po1ymorphisms. Cytogenet. Cell Genet. 13:536-550.

Miller DA, Tantravahi R, Dev VO, Miller OJ. 1977. Frequency of association of human chromosomes is correlated with amount of Ag-staining of the nucleolus organizer region. Amer J Hum Genet. 29:490-502.

Miller OJ, Miller DA, Alderdice PW, Dev VC, Grewal MS. 1971. Quinacrine fluorescent karyotypes of human diploid and heteroploid cell lines. Cytogenetics. 10:338.

Miller OJ, Miller DA, Warburton O. 1973. Application of new staining techniques to the study of human chromosomes. In: Progress in Medical Genetics. Steinberg A.G. and Bern A.G.(eds). Grune and Stratton. New York and London, 3.

Miroshnichenko LD, Pelipas VE. 1991. Regional differences in the prevalence of alcoholism in the USSR. Questions of Addiction. 3:29-32 (in Russian).

Mirre G, Stahl A. 1978. Ultrastructure and activity of the nucleolar organizer in the mouse oocyte during meiotic prophase. J Cell Sci. 31:79-100.

Mitchell AR, Bostock CJ. 1977. Restriction enzyme studies on human satellite III DNA. Heredity. 38:270.

Montague CT. 1997. Congenital leptin deficiency is associated with severe early-onset obesity in humans. Nature. 387:903-908.

Moor R. 1964. Alcoholism in Japan. Quart J Stud Alcohol. 25:142-150.

Mortland FW, de Bruin PP, Smith NH. 1954. Spectrophotometric studies on the interactions of nucleic acids with amino acridines and other basic dyes. Exp Cell Res. 7:201-214.

Moscetti G, Petriaggi M, Barbarossa CG, Tiberti S. 1971. Fluorescence staining method for the morphological and structural study of human chromosomes. Humangenetik. 12(1):56-58.

Muldal S, Ockey CH. 1961. Muscular dystrophy and deletion of Y chromosome. Lancet. ii: 601.

Müller H, Klinger HP. 1976. Chromosome polymorphism in a human newborn population. Part I. In: Chromosomes Today. Pearson P.L. and Lewis K.R. (eds.). Jerusalem: Israel Universities Press. New York: Wiley, p. 249.

Müller HJ, Klinger HP, Glasser M. 1975. Chromosome polymorphism in a human newborn population. II. Potentials of polymorphic chromosome variants for characterizing the idiograms of an individual. Cytogenet Cell Genet. 15:239-255.

Nakagome Y, Kitagawa Z, Linnmak K. et al. 1977. Pitfalls in the use of chromosome variants for paternity dispute cases. Hum Genet. 37:255-260.

Narberhaus F, Waldminghaus T, Chowdhury S. 2006. RNA thermometers. FEMS Microbiology Reviews. 30(1):3-16.

Natarajan AT, Ahnström G. 1972. Induced chromosomal aberration and heterochromatin. In: Modern aspects of cytogenetics: Constitutive heterochromatin in man. pp. 216-223.

Nazarenko SA. 1987. Age dynamics of human chromosome fluorescent polymorphism. Cytol Genet. 21:183-186.

Nielsen J. 1978. Large Y chromosome (Yq+) and increased risk of abortion. Clin Genet. 13:415-416.

Nielsen J, Friedrich U. 1972. Lenth of the Y chromosome in criminal males. Clin Genet. 3:281.

Nielsen J, Nortland E. 1975. Length of the Y chromosome and activity in boys. Clin Genet. 8:291.

Nielsen J, Friedrich U, Hreidarsson AB, Zeuthen E. 1974. Frequency and segregation of 16qh+. Clin Genet. 5:316-321.

Noel B, Emerit I, Luciani JM, Quack B. 1971. A familial Y/autosome translocation in man. Clin Genet. 2:1.

O'Brien RL, Olenick JG, Hahn FE. 1966. Reactions of quinine, chloriquine, and quinacrine with DNA and their effects on the DNA and RNA polymerase reactions. Proc Nat Acad Sci USA. 55:1511.

O'Riordan ML, Robinson JA, Buckton KE, Evans HJ. 1971. Distinguishing between the chromosomes involved in Down's syndrome (trisomy 21) and chronic myeloid leukemia (Ph ') by fluorescence. Nature. 230:167-168.

Ohno S. 1967. Sex chromosomes and sex-linked genes. Springer Verlag. Berlin, Heidelberg, New York, 1967.

Orgel LE, Crick FHC. 1980. Selfish DNA: the ultimate parasite. Nature. 284:604-607.

Paddy MR, Belmont AS, Saumweber H, Agard DA, Sedat JW. 1990. Interphase nuclear envelope lamins from a discontinuous network that interacts with only a fraction of the chromatin in the nuclear periphery. Cell. 62:89-106.

Pardue ML, Gall JG. 1970. Chromosomal localization of mouse satellite DNA. Science. 170:1356-1358.

Paris Conference, 1971, and Supplement 1975. Standartization in human cytogenetics. Birth Defects: Original Article Series, XI, 1-84. The National Foundation, New York.

Park J. 1976. Human C- band polymorphisms: frequency and distribution in Orientals. Ph. D. Thesis, Indiana University.

Parry DM, Sandler L. 1974. The genetic identification of a heterochromatic segments on the X chromosome of *Drosophila melanogaster*. Genetics. 77:535-539.

Passarge E. 1979. Emil Heitz and the concept of heterochromatin: longitudinal chromosome differentiation was recognized fifty years ago. Am J Hum Genet. 31:106-115.

Patau K. 1960. The identification of individual chromosomes, especially in man. Am J Hum Genet. 12:250-276.

Pathak S, Würster Hill DH. 1977. Distribution of constitutive heterochromatin in Carnivores. Cytogenet Cell Genet. 18:245-254.

Pathak S, Hsu TC, Arrighi FE. 1973. Chromosomes of peromyscus (Rodentia, Cricetidae). IV. The role of heterochromatin in karyotypic evolution. Cytogenet Cell Genet. 12:315-326.

Patil SR, Lubs AA. 1977a. Classification of qh regions in human chromosomes 1, 9 and 16 by C banding. Hum Genet. 38:35-38.

Patil SR, Lubs HA. 1977b. A possible association of long Y-chromosomes and fetal loss. Hum Genet. 35:233-235.

Peacocke AB, Scarrett JNH. 1956. The interaction of aminoacridines with nucleic acids. Trans Faraday Soc. 52:261.

Pearson P. 1972. The use of new staining technique for human chromosome identification. J Med Genet. 9(3):264-275.

Pearson PL. 1973. The uniqueness of the human karyotype. In: Chromosome identification techniques and application in biology and medicine. Caspersson T. and Zech L. (eds.). New York, London. Academic Press, p. 145.

Pearson PL. 1977. Pattern of bands, polymorphism and evolution of primates. In Molecular structure of human chromosomes. Yunis J.J. (Ed). Acad. Press. p. 267.

Pearson PL, Bobrow M, Vosa GG. 1970. Technique for identifying Y chromosome in human interphase nuclei. Nature. 226:78-80.

Pearson PL, Bobrow M, Vosa CG, Barlow PN. 1971. Quinacrine fluorescence in mammalian chromosomes. Nature. 231:326-329.

Pelleymounter MA, Gullen MJ, Baker MB, Hecht R, Winters D, Boon T, Collins F. 1995. Effects of the obese gene product on body weight regulation in OB/OB mice. Science. 269:540-543.

Pera F. 1968. Dauer der DNS-Replication von Eunnd Heterochromatin bei Microtus agrestis. Chromosoma. 25:21-29.

Phillips RB. 1977. Inheritance of Q and C-band polymorphisms. Can J Genet Cytol. 19:405-415.

Phillips RB. 1980. New C band markers of human chromosomes. C band position variants. J Med Genet. 17:380-385.

Podugolnikova OA, Sushanlo HM, Parfenova IV, Prokofyeva-Belgovskaya AA. 1979. The quantitative analysis of polymorphism on human chromosomes 1, 9, 16 and Y. II. Comparison of the C- segments in male and female individuals (group characteristics). Hum Genet. 49:251-260.

Priest JH, Peakman DC, Patil SR, Robinson A. 1970. Significance of chromosome 17ps+ in three generations of a family. J Med Genet. 7:142.

Prokofyeva-Belgovskaya AA. 1986. Heterochromatin Regions of Chromosomes (Russian). Moscow: Nauka

Rae PMM, Franke WW. 1972. The interphase distribution of satellite DNA-containing heterochromatin in mouse nuclei. Chromosoma. 39:443-456.

Raposa T, Natarajan AT. 1974. Fluorescence banding pattern of human and mouse chromosomes with benzimidazole derivative (Hoechst 33258). Humangenetik. 21:221-226.

Ricci N., Gastoldi G.L., Dallapicoola B. and Baserga A. 1971. Small drumsticks and long Y chromosomes. Br. Med. J., 1: 346.

Robinson IA, Newton M. 1977. A fluorescence polymorphism associated with Down's syndrome. J Med Genet. 14:40-45.

Robinson JA, Buckton KE, Spowart G, et al. 1976. The segregation of human chromosome polymorphisms. Ann Hum Genet. 40:113-121.

Rossi SL. 1985. Polymorphismos chromosomicos: metodologia y application en diagnostic prenatal. Theses Profecional. Mexico.

Rowley PT, Marshall R, Ellis JR. 1963. A genetical and cytological study of repeated spontaneous abortion. Ann Hum Genet. 27:87.

Ruthner U, Golob E. 1974. 45, X/45, X, acc (?Yp) + /46, X r(Y) in a phenotypically normal newborn male. Humangenetik. 22:177.

Sadoni N, Langer S, Fauth C. et al. 1999. Nuclear organization of mammalian genomes: polar chromosome territories build up functionally distinct higher order compartments. J Cell Biol. 146:1211-1226.

Sandberg AA. (Ed). 1985. Progress and Topics in Cytogenetics. The Y Chromosome. Part A and B. Alan R. Liss, Inc., New York.

Sanders PF, Pavan C. 1972. Heterochromatin in development of normal and infected cells. In: Cell differentiation. Ed. R. Harris et al., Copenhagen. 162-175.

Sarich VM, 1971. A Molecular Approach to the Study of Human Origins, in Background for Man, ed. P. Dolhinow and V. M. Sarich (Boston: Little, Brown). 60-81.

Scheinfeldt LB, Soi S, Thompson S, et al. 2012. Genetic adaptation to high altitude in the Ethiopian highlands.Genome Biology. 13:R1.

Schmid W. 1963. DNA replication patterns of human chromosomes. Cytogenetics 36:175.

Schmid W. 1967. Heterochromatin in mammals. Arch. Julius K1aus-Stiftung Vererb. 42:1-60.

Schmid M, Vogel W, Krone W. 1975. Attraction between centric heterochromatin of human chromosomes. Cytogenet. Cell Genet. 15:66-80.

Schnedl W. 1971. Unterachiedliche Fluorescenc der beiden homologen Chromosomen No. 3 heim Menschen. Humangenetik. 12:59-63.

Schnedl W. 1972. Giemsa banding, quinacrine fluorescence and DNA-replication in chromosomes of cattle (Bos taurus). Chromosoma. 38:319-328.

Schnedl W, Brietenbach M, Mikelsaar AB, Stranzinger C. 1977. Mithramycin and DIPI: a pair of fluorochromes specific for GC- and AT-rich DNA respectively. Hum Genet. 36:299-305.

Schreck R, Warburton D, Miller OJ. et al. 1973. Chromosome structure as revealed by a combined chemical and immunochemical procedure. Proc Natl Acad Sci USA. 70(3):804-807.

Schull WJ, Rothhammer F. 1977a. A multinational Andean genetic and health programme: A study of adaptation to the hypoxia of altitude. In: Genetic and Nongenetic Components in Physiological Variability. Weiner JS (Ed). London: Society for the study of Human Biology. 17:139-169.

Schull WJ, Rothhammer F. 1977b. Ananlytic methods for genetic and adaptational studies. In: Origins and Affinities of the First Americans. Launghlin W.S. (Ed). New York: Gustav Fisher.

Schwartz S, Cohen MM. 1985. Y chromosome length: clinical manifestation and reproductive capacity. In: Progress and Topics in Cytogenetics. Vol. 6. Pt. B. Sandberg A.A. (Ed). Alan R. Liss Inc., N.Y. 27-52.

Schwartz MW, Woods SC, Porte DJr, Seeley RJ, Baskin D.G. 2000. Central nervous system control of food intake. Nature. 404:661-671.

Schwinger E, Wild P. 1974. Length of the Y chromosomal and antisocial behavior? Hum Genet. 22:67.

Schwinger E, Wehner H. 1976. Frequency of chromosomal fluorescence polymorphism in normal persons and in clinical patients with diagnosed chromosome aberrations. Hum Genet. 32:115-119.

Selander RK. 1973. Interaction of quinacrine mustard with mononucleotides and polynucleotides. Biochem J. 131:749-755.

Selander RK, de la Chapelle A. 1973. The fluorescence of quinacrine mustard with nucleic acids. Nature New Biol. 245:240-244.

Seuanez H, Fletcher J, Evans HJ, Martin DE. 1976. A polymorhic structural rearrangement in the chromosomes of two populations of orangutan. Cytogenet Cell Genet. 17:317-326.

Sibley CG, Alquist JE. 1984. The Phylogeny of the Hominid Primates as Indicated by DNA-DNA Hybridization. J Mol Evol. 20:2-15.

Sieberg JW, Vogel W, Herp H, Bolze H, Dittrich A. 1973. Structural aberrations of the chromosome and the corresponding phenotype. Report of a case with the karyotype 45, X/46, X, i (Yp). Humangenetik. 19:57.

Sieger M, Pera F, Schwarzacher HG. 1970. Genetic inactivity of heterochromatin and heteropicnosis in Microtus agrestis. Chromosoma. 29:349-364.

Simonson TS, Yang Y, Huff CD, et al. 2010. Genetic evidence for high altitude adaptation in Tibet. Science. 329:72-75.

Skawinski W, Parcheta B. 1984. Polymorphism of the human Y chromosome: the evaluation of the correlation between the DNA content and the size of the heterochromatin and euchromatin. Clin Genet. 25:125-30.

Solomon IL, Hamm CW, Green OC. 1964. Chromosome studies on testicular tissue cultures and blood leukocytes of a male previously reported to have no Y chromosome. New Engl J Med. 271:586.

Soudek D. 1977. Possible risk factors associated with enlarged constitutive heterochromatin. Fifth Inter. Cont. on Birth Defeets, 21-27. August. pp. 116-117.

Soudek D, Laraya P. 1976. C and Q bands in long arm of Y chromosomes: are they identical? Hum Genet. 32:339-341.

Soudek D, Sroka H. 1978. Inversion of "fluorescent" segment in chromosome 3: A polymorphic trait. Hum Genet. 44:109-115.

Soudek D, Langmuir V, Steward BJ. 1973. Variation in the nonfluorescent segment of long Y chromosome. Humangenetik. 18:285.

Soudek D, O'Shaunghnesky, Laraya P, McCreary BD. 1974. Pericentric inversion of "fluorescent" segment in chromosome No 3. Humangenetik. 22:343-346.

Sparvoli E, Galli MG, Chiatante D, Sgorbati S. 1977. Perinuclear DNA and heterochromatin. Exp Cell Res. 110:315-321.

Stahl A, Hartung MI. 1981. L'heterochromatine. Ann Genet. 24:69-77.

Stahl A, Luciani JM, Devictor M, Capodano AM. 1975. Constitutive heterochromatin and micronucleoli in the human oocyte at the diplotene stage. Humangenetik 26:315-327.

Stanyon R, Studer M, Dragone A, De Benedicts G, Brancati C. 1988. Population cytogenetics of Albanians in the province of Cosenza (Italy): frequency of Q and C band variants. Int J Anthropol. 3(1):14-29.

Stunkard AJ, Harris JR, Pedersen NI, McClearn GA. 1990. The body-mass index of twins who have been reared apart. N Engl J Med. 322:1483-1487.

Sumner AJ. 1972. A simple technique for demonstration centromeric heterochromatin. Exp Cell Res. 83:438-442.

Sumner AT, Evans HS, Buckland KA. 1971. New technique for distinguishing between human chromosomes. Nature (Lond.) New Biol. 232:31.

Swanson CP, Merz T, Young WJ. 1967. Cytogenetics. Prentice-Hall, Inc. Englewood Cliffs, New Jersey.

Tanabe HS, Muller M, Neusser M, van Hase J, Caleagno E, Cremer M, Solovey I, Cremer C, Cremer T. 2002. Evolutionary conservation of chromosome territory arrangement in cell nuclei from higher primates. Proc Natl Acad Sci USA. 99:4424-4429.

Tada T, Takagi N, Adler ID. 1993. Parental imprinting on the mouse X chromosome: effects on the early development of X0, XXY and XXX embryos. Genet Res. 62(2):139-48.

Takagi N, Sugawara O, Sasaki M. 1982. Regional and temporal changes in the pattern of X-chromosome replication during the early post-implantation development of the female mouse. Chromosoma, 85(2):275-286.

Tanguay R, Gagne R, Laberge C. 1971. Differential staining patterns of heterochromatin in man. Nature New Biol. 232(27):29-30.

Taylor JH. 1960. Asynchronous duplication of chromosomes in cultured cells of Chinese hamster. J Biophys Biochem Cytol. 7:455-464.

Tharapel AT, Summit RL. 1978. Minor chromosome variations and selected homeomorphisms in 200 unclassifiable mentally retarded patients and 200 normal controls. Hum Genet. 41:121-130.

Therkelsen AJ, Lamm LV, Henningsen K. 1967. Heritable morphological difference between the two chromosomes No.16 in humans. Hereditas. 57:149.

Therman E. 1986. Human Chromosomes: Strucnure, Behavior, Effects. Springer-Verlag, New York, Berlin, Heidelberg, Tokyo.

Tishler PV, Lamborot-Manzur M, Atteins L. 1972. Polymorphisms of the human Y chromosome; Fluorescence microscopic studies on the sites of morphologic variation. Clin Genet. 3:116.

Tjio JH, Levan A. 1956. The chromosome number of man. Hereditas. 42:1-6.

Tupitsina LP, Stobetsky VI. 1980. Segregation of Q-polymorphic variants of human chromosomes. Genetics. 16:727 (in Russian).

Xu S, Li S, Yang Y, et al, 2011. Agenome-wide search for signals of high-altitude adaptation in Tibetans. Mol Biol Evol. 28:1003-1011.

Vagner-Capodano AM, Pinna-Delgrossi MH, Stahl A. 1977. Localisation of nucleolar organizers in the interphase nucleus of human fibroblasts. Biol Cell. 29:209-210.

Vaisse C, Clement K, Guy-Grand B, Froguel P. 1998. A frame shift mutation in human MC4R is associated with a dominant form of obesity. Nature Genet. 20:113-114.

van der Hagen CB, Berg K. 1970. Studies of human chromosomes by DNA-binding fluorochromes. I. The normal chromosome complement. Clin Genet. 1:263-271.

van Holde KE. 1988. Chromatin Springer-Verlag, New York.

van Driel R, Wansink DG, van Steensel B, Grande W, de Jong L. 1995. Nuclear domains and the nuclear matrix. Int Rev Cytol. 162A:151-189.

van Dyke DL, Palmer CG, Nance WE, Pao-Lo Yu. 1977. Chromosome polymorphism and twin zygosity. Am J Hum Genet. 29:431-447.

van Holde KE. 1988. Chromatin. Springer-Verlag, New York.

Verma RS. (Ed). 1988. Heterochromatin. Molecular and Structural Aspects. Cambridge Univ. Press. Cambridge, New York, Sydney.

Verma RS, Dosik H. 1976. An improved method of photographing human fluorescent chromosomes. J Microsc. 108:339.

Verma RS, Dosik H. 1980a. Human chromosomai heteromorphisms: nature and clinical significance. Int Rev Cytol. 62:361-383.

Verma RS, Dosik H. 1980b. Human chromosomal heteromorphisms in American blacks. I. Structural variability of chromosome 3. J Hered. 71:441.

Verma RS, Dosik H. 1982. Human chromosomal heteromorphisms in American blacks. VII. Correlation between height and the size of the Y chromosome. Jpn J Hum Genet. 27:43.

Verma RS, Dosik H, Lubs HA. 1978a. Size and pericentric inversion heteromorphisms of secondary constriction regions (h) of chromosomes 1, 9 and 16 as detected by CBG technique in Caucasians: classification, frequencies, and incidence. Amer J Med Genet. 2:331-339.

Verma RS, Dosik H, Scharf T, Lubs HA. 1978b. Length heteromorphisms of fluorescent (f) and non-fluorescent (nf) segments of human Y chromosome: classification, frequencies, and incidence in normal Caucasians. J Med Genet. 15:277.

Verma RS, Benjamin C, Rodriguez J, Dosik H. 1981. Population heteromorphisms of Ag-stained nucleolus organizer regions (NORs) in the acrocentric chromosomes of East Indians. Hum Genet. 59:412-415.

Verma RS, Huq A, Dosik H. 1983. Racial variation of a non-fluorescent segment of the Y-chromosome in East Indians. J Med Genet. 20:102.

Vogt P. 1990. Potential genetic functions of tandem repeated DNA sequence blocks in the human genome are based on a highly conserved "chromatin folding code." Hum Genet. 84:301-336.

Vosa CG. 1971. The quinacrine fluorescence patterns of the chromosomes of *Allium carinatum*. Chromosoma. 33:382-385.

Wachtel SS, Koo GC, Breg S. et al. 1975. Expression of H-Y antigen in the human males with two Y chromosomes. New Engl J Med. 293:1070.

Walker PMB. 1971. Origin of satellite DNA. Nature 229:306-308.

Warburton D, Firschein IL, Miller DA, Warburton F. 1973. Karyotype of the chimpanzee Pan troglodytes based on measurements and banding pattern: comparison to the human karyotype. Cytogenet Cell Genet. 12:452-461.

Warden CH, Fisher JS, Shoemaker SM. 1995. Identification of four chromosomal loci determining obesity in a multifactorial mouse model. J Clin Invest. 95:1545-1552.

Waring MJ. 1965. Complex formation between ethidium bromide and nucleic acids. J Mol Biol. 13:269-282.

Wehlstrom J. 1971. Are variation in the length of Y chromosome due to structural changes? Hereditas. 69:125.

Wehlstrom J, de la Chapelle A. 1963. Elongation as the possible mechanisms of the origin of large human chromosomes. Hereditas. 50:345.

Weisblum B. 1973. Why centric regions of quinacrine-treated mouse chromosomes show diminished fluorescence. Nature. 246:150-151.

Weisblum B. 1974. Fluorescent probes of chromosomal DNA structure: three classes of acridines. In: Chromosome structure and function. Cold Spring Harbor Symp. Quant. Biol. 38:441-449.

Weisblum B, de Haseth PL. 1972. Quinacrine, a chromosome stain specific for deoxyadenylate-deoxythymidilate-rich regions in DNA. Proc Nat Acad Sci USA. 69:629-632.

West DB, Goudey-Lefevre J, York B, Truett GE. 1994. Dietary obesity linked to genetic loci on chromosomes 9 and 15 in a polygenic mouse model. J Clin Invest. 94:1410-1416.

Westlike JR, Robertson R, Leddet I. et al. 1983. Y-chromosome length related to fetal loss. Clin Genet. 24: 413-419.

Yamada K. and Hasegawa T. 1978. Types and frequencies of Q-variant chromosomes in a Japanese population. Hum. Genet. 44: 89-98.

Yamada K, Ohta M, Yoshimura K, Hasekura H. 1981. A possible association of Y chromosome heterochromatin with stature. Hum Genet. 58:268.

Yasmineh WG, Yunis JJ. 1973. The repeated DNA sequences of Microtinae. I. *Microtus agrestis, Microtus pennsylvanicus* and *Ellabius lutesceus.* Exp Cell Res. 81:432-446.

Yi X, Liang Y, Huerta-Sanchez E, et al. 2010. Sequencing of 50 human exomes reveals adaptation to high altitude. Science. 329:75-78.

Yosida TH. 1975. Diminution of heterochromatic C-bands in relation to the differentiation of Rattus species. Proc Jap Acad Sci. 51:659.

Yunis SS, Yasmineh WG. 1971. Heterochromatin, satellite DNA and cell function. Science. 174:1200-1209.

Yunis JJ, Prakash Y. 1982. The origin of man: a chromosomal pictorial legacy. Science. 215:1525-1530.

Yunis JJ, Rolda L, Yasmineh WG, Lee JC. 1971. Staining of satellite DNA in metaphase chromosome. Nature. 231:532.

Zakharov AF, Tsoneva MT. 1982. Chromosome polymorphism in human populations. In: Prospects for Medical Genetics Medicine, Moscow (in Russian).

Zakharov AF, Benyush VA, Kuleshov NP, Baranovskaya LI. 1982. Human chromosomes. Atlas. The medicine. Moscow (in Russian).

Zech L. 1969. Investigation of metaphase chromosomes with DNA-binding fluorochromes. Exp Cell Res. 58:463.

Zhang Y, Proenca R, Maffei M, et al. 1994. Positional cloning of the mouse obese gene and its human homologue. Nature. 372:425-432.

Zhang L-F, Huynh KD, Lee JT. 2007. Perinucleolar targeting of the inactive X during S phase: evidence for a role in the maintenance of silencing. Cell. 129(4):693-706.

Zink D, Cremer T. 1998. Cell nucleus: chromosome dynamics in nuclei of living cells. Curr Biol. 8:R321-R324.

Zink D, Cremer T, Saffrich R. et al. 1998. Structure and dynamics of human interphase chromosome territories in vivo. Hum Genet. 102(2):241-51.

INDEX